Development of a Strategic Planning Process

The mission of the Awwa Research Foundation (AwwaRF) is to advance the science of water to improve the quality of life. Funded primarily through annual subscription payments from over 1,000 utilities, consulting firms, and manufacturers in North America and abroad, AwwaRF sponsors research on all aspects of drinking water, including supply and resources, treatment, monitoring and analysis, distribution, management, and health effects.

From its headquarters in Denver, Colorado, the AwwaRF staff directs and supports the efforts of over 700 volunteers, who are the heart of the research program. These volunteers, serving on various boards and committees, use their expertise to select and monitor research studies to benefit the entire drinking water community.

Research findings are disseminated through a number of technology transfer activities, including research reports, conferences, videotape summaries, and periodicals.

Development of a Strategic Planning Process

Prepared by:

CH2MHILL

Sponsored by:
Awwa Research Foundation
6666 West Quincy Avenue, Denver, CO 80235-3098

Published by:

and

DISCLAIMER

This study was jointly funded by the Awwa Research Foundation (AwwaRF) and the Honolulu Board of Water Supply. AwwaRF and the Honolulu Board of Water Supply assume no responsibility for the content of the research study reported in this publication or for the opinions or statements of fact expressed in the report. The mention of trade names for commercial products does not represent or imply the approval or endorsement of AwwaRF or the Honolulu Board of Water Supply. This report is presented solely for informational purposes.

Library of Congress Cataloging-in-Publication Data

Development of a strategic planning process: tailored collaboration project: final report.
p. cm.
Includes bibliographical references.
"Prepared for AWWA Research Foundation and Honolulu Board of Water Supply."
ISBN 1-58321-284-1
1. Water utilities—United States —Planning. 2. Strategic planning—United States. 3. Water utilities—Hawaii—Honolulu—Planning—Case studies. I. AWWA Research Foundation. II. American Water Works Association. III. Honolulu (Hawaii). Board of Water Supply.

HD4461.D48 2003
363.6'1'0684—dc21 2 00 3056064

ISBN 1-58321-284-1

CONTENTS

FOREWORD

The Awwa Research Foundation is a nonprofit corporation that is dedicated to the implementation of a research effort to help utilities respond to regulatory requirements and traditional high-priority concerns of the industry. The research agenda is developed through a process of consultation with subscribers and drinking water professionals. Under the umbrella of a Strategic Research Plan, the Research Advisory Council prioritizes the suggested project based upon current and future needs, applicability, and past work; the recommendations are forwarded to the Board of Trustees for final selection. The foundation also sponsors research projects through an unsolicited proposal process; the Collaborative Research, Research Applications, and Tailored Collaboration programs; and various efforts with organizations such as the U.S. Environmental Protection Agency, the U.S. Bureau of Reclamation, and the Association of California Water Agencies.

This publication is a result of one of these sponsored studies, and it is hoped that its findings will be applied in communities throughout the world. The following report serves not only as a means of communicating the results of the water industry's centralized research program but also as a tool to enlist the further support of the nonmember utilities and individuals.

Projects are managed closely from their inception to the final report by the foundation's staff and large cadre of volunteers who willingly contribute their time and expertise. The foundation serves a planning and management function and awards contracts to other institutions such as water utilities, universities, and engineering firms. The funding for this research effort comes primarily from the Subscription Program, through which water utilities subscribe to the research program and make an annual payment proportionate to the volume of water they deliver and consultants and manufacturers subscribe based on their annual billings. The program offers a cost-effective and fair method for funding research in the public interest.

A broad spectrum of water supply issues is addressed by the foundation's research agenda: resources, treatment and operations, distribution and storage, water quality and analysis, toxicology, economics, and management. The ultimate purpose of the coordinated effort is to assist water suppliers to provide the highest possible quality of water economically and reliably. The true benefits are realized when the results are implemented at the utility level. The foundation's trustees are pleased to offer this publication as a contribution toward that end.

Edmund G. Archuleta, P.E.
Chair, Board of Trustees
Awwa Research Foundation

James F. Manwaring, P.E.
Executive Director
Awwa Research Foundation

ACKNOWLEDGMENTS

The Awwa Research Foundation and CH2M HILL wish to acknowledge the generous support of the Honolulu Board of Water Supply (BWS) and the members of our Utility Management Applications Board in completing this project, and particularly the support and guidance of Walter Bishop and Kurt Ladensack of Contra Costa Water District, Concord, California. Members of the Applications Board included:

- Walter Bishop, Contra Costa Water District, Concord, Calif.
- Julius Ciaccia, Cleveland Division of Water, Cleveland, Ohio.
- Jon Eckenbach, JEA, Jacksonville, Fla.
- Diana Gale, Seattle Public Utilities, Seattle, Wash.
- John Huber, Louisville Water Company, Louisville, Ky.
- Adam Kramer, Minneapolis Public Works Department, Minneapolis, Minn.
- Mark Premo, Anchorage Water and Wastewater Utility, Anchorage, Alaska
- Mark Rosenberger, City of Portland Water Bureau, Portland, Ore.

Representatives from the Honolulu Board of Water Supply whose support was instrumental throughout this project included Cliff Jamile, Donna Kiyosaki, and Susan Uyesugi; the entire BWS management team contributed patiently and materially to the project's completion.

Chapter 1: Introduction

About this project

Water utility management has become increasingly dynamic, competitive, and complex due to a variety of influences shaping our industry. Traditional approaches to project and service delivery are being challenged, and preservation of the status quo is untenable. In particular, water utilities must continue to safeguard public health while facing multiple challenges such as constrained resources, aging infrastructure, and new forms of competition.

To effectively plan and guide their ongoing transformation in a competitive marketplace, utilities must adopt a strategic business perspective; specifically, utilities can enhance their strategic position through decision-making techniques that characterize successful business organizations.

Recognizing this industry need, the Awwa Research Foundation (Research Foundation) undertook a tailored collaboration project, entitled *Development of a Strategic Planning Process*, together with the Honolulu Board of Water Supply (BWS) and CH2M HILL. The result of this project is a template for strategic planning in water utilities – one that integrates ongoing discrete planning activities including asset management, capital improvement planning, integrated resource planning, public communications, and financial planning.

Attributes of the strategic planning process:

- Simple, streamlined, understandable, and accessible
- Provides a bridge between strategic goals and annual budget
- Adaptable to an individual utility's organizational culture
- Embedded in the way utilities do business
- Inclusive of stakeholder groups
- Valuable to utilities whether undertaken as a whole or in individual steps
- Finite, though updated annually

Project objective and goals

The primary project objective was to develop, and demonstrate the application of, a strategic planning process for water utilities. Project goals included:

- Develop and demonstrate a flexible, adaptable process for strategic planning in water utilities.
- Demonstrate integration of the strategic planning process into annual utility planning activities.
- Test the applicability of private business approaches (specifically, portfolio management) to strategic planning in water utilities.

Products of this project include:

- The strategic planning process and this summary report
- A draft strategic plan for the Honolulu Board of Water Supply (included as Appendix A to this report)

Why strategic planning?

Water utility management has increasingly required an entrepreneurial outlook in recent years, as the institutional framework for service delivery has moved from monopoly to a market focus in many areas. This shift has been signaled by several developments including:

- **Customer and stakeholder expectations** have risen in proportion to their access to information – and the focus on government efficiency.
- **Traditionally "protected" functions** have been challenged by the use of alternative methods for project and service delivery.
- **The lack of federal and state support** for infrastructure investments, together with limited local government financing capacities, have intensified the pressure on utilities facing major capital needs.

When this market evolution began, the industry took a largely reactive approach to the effects of heightened competition. This fostered cost-saving advances, but it also introduced the possibility that market advantages may be sacrificed in the name of competitiveness.

Exacerbating the situation, the dynamic nature of the utility marketplace has been cast in sharp relief by increasing threats of global terrorism, heightened scrutiny of accounting practices, and the continued general decline of the global economy.

The true market challenge is not simply a matter of closing competitiveness gaps or reacting to perceived threats. Rather, water utilities must understand the spectrum of potential market conditions and develop strategies that

- Protect their ability to continue delivering reliable, cost-efficient service, and
- Provide opportunities for organizational growth and development.

This challenge entails deciding how to invest limited resources to achieve the greatest desired returns, within acceptable levels of risk.

Applicability to water utilities

Faced with this challenge, water utilities can draw on approaches that private business entities use to conduct strategic planning. It is important to begin with a clear, common understanding of what a strategic plan is: *a tool for making resource allocation decisions among core business functions, and investments in expanding and/or diversifying business functions, in a way that positions the utility to* ***increase value to customers/shareholders.***

Definition of a strategic plan:
A tool for making resource allocation decisions among core business functions, and investments in expanding and/or diversifying business functions, in a way that positions the utility to ***increase value to customers/ shareholders.***

The concept of value to shareholders is quite different for public utilities than for private entities. In the private arena, value to shareholders is easily measured in primarily monetary terms. For public utilities, however, shareholders or stakeholders bring to the table an extraordinary range of competing priorities – including, first and foremost, the provision of safe drinking water, but complicated by environmental, regulatory, political, economic, community, and social agendas. Further clouding the situation, utilities conduct multiple planning activities that are often disjointed, with no reliable way to measure how each one is *truly* contributing to achieving the utility's fundamental organizational goals.

Strategic planning provides a potential solution, bringing planning activities together under a common framework that links all planning and budgeting decisions to the utility's fundamental goals.

Issues and concerns

The Utility Management Applications Board identified a number of issues and concerns relative to strategic planning in water utilities that substantially affected the focus and outcome of this project. While many of these issues are addressed at least in part herein, others will be addressed in the Research Foundation's planned FY 2002-03 project, *Strategic Planning and Organizational Development for Water Utilities*. This follow-on effort is intended to delineate, with detailed guidance, a strategic planning process applicable to water utilities in general, including appropriate tools and techniques from private industry.

"It's not realistic to expect the unequivocal commitment of your policy board to the strategic planning process and outcome at the outset. A commitment to ***constructively participate in strategic planning*** is a goal that can fit governing officials' need to be flexible in response to changing circumstances."

Walter Bishop
Contra Costa Water District

In brief, the top issues and concerns raised by Applications Board members included:

- **Strategic planning requires a substantial commitment** from executive leadership and policymakers (governing boards, councils, elected officials) to the planning process and its potential outcome. In the absence of this commitment, a strategic plan is likely to create unwarranted staff expectations and confusion between strategic planning, other planning initiatives, and annual budgeting.
- **Utilities need to set expectations** for policy boards and stakeholders about what strategic planning *is* and *isn't*. Utilities must be realistic about the implementability of outcomes from strategic planning – and take care not to raise over-inflated expectations of change.
- **Business planning terminology** may be daunting to utility staff. Some education in business planning techniques is recommended for utility management team.
- **Strategic planning must be iterative**, have long-range value, and improve with age. Utilities with immature planning processes (and thereby insufficient data to evaluate investment options) will end up "planning to plan" in initial strategic planning cycles.

- **Strong leadership commitment** to participation in and implementation of the process is essential for success.
- **Varying levels of detail and analytical complexity** in strategic planning are needed for different utility sizes and circumstances. The level of detail should be matched to appropriate target audiences within the utility; executive managers and policy boards will not have patience with details.
- **Water utilities need to address barriers to their expanded participation in the competitive marketplace.** The changes in water utility market conditions have not been accompanied by parallel changes in the institutional and legal frameworks within which many utilities operate. "Entrepreneurial" water utilities will need to plan for redress of institutional and legal barriers to their participation in various forms of enterprise.

Changes in the water utility marketplace have not been accompanied by parallel changes in institutional and legal frameworks within which many utilities operate. ***"Entrepreneurial" water utilities must address institutional and legal barriers*** to their participation in various forms of enterprise.

- **Scenario planning** is a useful tool for analyzing business strategies in an uncertain environment, because it requires consideration of a broad range of market conditions and possible influences.
- **In identifying investment options**, utilities must be careful to clarify the difference between core investments (those investments required to keep the utility in operation) and strategic investments (discretionary investments to enhance value of services, expand service offerings, or generate new revenue sources).
- **Utilities need further guidance on prioritizing** "soft" (people, knowledge, communication) vs. "hard" asset investments.
- **A business planning tool**, or template, is needed (see Chapter 5, Step 3; and Appendix B) to drive goal attainment down into the utility and into its various planning activities.
- **Customers and community are separate stakeholder groups** and should be addressed individually in evaluating investment options and in communicating about the strategic planning process.
- **Strategic planning must be accompanied by a communication plan**, largely focused on utility staff, to engender a cultural shift to a goals-focused organization.

Getting started

What to expect from strategic planning

Expectations of strategic planning must be considered in relation to the existence, validity, and maturity of asset master plans and long-range financial plans. One of the first steps in the recommended strategic planning process is an evaluation of existing planning processes. If a utility does not have rigorous, formal processes in place for master, capital, and financial planning, the utility is cautioned against making investment decisions of major consequence in the early years of strategic planning. For example, if the utility conducts little or no medium- or long-term planning for water supply, physical system assets, or information systems, the ability to generate a realistic strategic plan is compromised. For utilities in this situation, strategic planning can be approached initially as a "plan to plan." Many of the strategic investments such a utility may consider likely will include implementation of more formal planning processes and tools, along with the data collection efforts needed to support them.

Fundamentally, strategic planning is an evolutionary process wherein the tools and techniques used, as well as participants' acumen, are improved over time. Accordingly, one may expect initial strategic planning efforts to largely reflect participants' intuition rather than analytical rigor and measurable results.

Is strategic planning right for your utility?

Virtually any utility will benefit from some level of strategic planning; however, there are circumstances under which strategic planning may not be appropriate, such as

1. The utility is experiencing crisis conditions requiring direct action that precludes the ability to invest time in strategic planning.
2. There is a lack of leadership commitment, time, or available resources to carry the process to conclusion.

Getting to a go/no-go decision

Before entering into strategic planning, utility management must consider a number of key questions to help structure the process for success and avoid pitfalls. Answers to these questions can help a utility tailor its strategic planning process to its "real-time" situation.

Important considerations include precedents for policymaker involvement and stakeholder acceptance, existing decision processes and protocols, funding constraints, available information from existing planning programs, and potential risks.

In addition, the utility must build a business case for strategic planning and present this case to its policy board to seek a go/no-go decision on entering the strategic planning process.

Recommended actions in getting to a go/no-go decision include:

- **Define the utility's drivers and expectations** for strategic planning. What are the chief challenges in managing the utility? Confirm that the utility is not considering the use of strategic planning to deal with a problem or situation not suited for high-level planning (such as crisis management).
- **Define the utility's baseline conditions**, as well as issues affecting operational success. Define the utility's strengths and weaknesses. Define how the utility compares with others against

basic performance metrics; ideally, the utility will have completed a competitiveness assessment from which to draw this information.

- **Consider key questions and policy issues** that influence strategic planning success. The step-by-step strategic planning instructions provided in Chapter 5 suggest typical policy issues and questions that will need to be addressed throughout the process. Review this information and consider the viability of getting the policy board's support throughout the strategic planning cycle.
- **Characterize the market situation** in the water utility industry, and in the utility's local/regional marketplace. Readily available sources, such as *Dawn of the Replacement Era: Reinvesting in Drinking Water Infrastructure* (AWWA, May 2001), catalog the challenges unique to the water utility industry and telegraph the need for strategic planning. Characterize the local and regional community – Is it economically strong or weakening? How strong are cost-control pressures? Who are the powerful influencers in the community? Describe the constraints and opportunities that the utility faces, both within and beyond its control.
- **Explore strategic plans from other utilities *and* private industry.** The Honolulu Board of Water Supply's draft 2001 strategic plan is provided as Appendix A. Other utilities will provide their strategic plans on request; some publish strategic plans on their websites. The Research Foundation's planned FY 2002-03 project, *Strategic Planning and Organizational Development for Water Utilities,* will take a closer look at the feasibility of adapting planning methods and tools from private industry.
- **Conduct scenario planning** and assess the business risk of maintaining the status quo. Scenario planning characterizes the business environment within which a utility will most likely operate over time and potential favorable or unfavorable changes that may affect business performance. Several scenarios may be developed to represent worst, expected, and best case conditions. By examining potential changes to its business environment, a utility may articulate a strategic direction and define risk-management methods that will buffer the organization against adverse business environment changes. Indicators or "triggers" that forewarn the utility of a change in the business environment may be identified so that steps can be taken to mitigate potential adverse consequences or take advantage of emerging opportunities.

  More details on scenario planning are provided in Chapter 5; another useful description of scenario planning – its purpose, outcomes, timeframe/level of effort, and a simple instruction process – are included in the Research Foundation's CD-ROM product entitled *Capital Planning Strategies.*

- **Estimate the schedule and level of effort** to prepare a strategic plan. Assess resource requirements and the utility's financial and managerial capability to complete the strategic planning process.

Key questions to consider in the go/no-go decision

- What changes might we expect in the business environment in the next 3 to 5 years?
- Are we concerned about protecting continued business viability?
- What role would the policy board want to play in developing a new strategy? Politically, what is our policy board willing to do to change utility strategic direction?
- What are the current constraints affecting the utility (financial, social, economic, regulatory, technological)? Which do we have the latitude to change?
- What are other players in our industry doing differently to meet customer needs?
- What possible business opportunities are on the horizon?
- What are the risks associated with changing – and with not changing? What is our level of risk tolerance?
- What are the influences that could cause us to fail in going forward?
- How does our utility measure up to others, in terms of standard performance metrics?

- **Consider the need for external support.** The ability of a utility's management team to seriously consider outside-the-box alternatives could be limited, in some utilities, by a historical focus on maintaining the status quo, and by relative inexperience with entrepreneurial initiatives. Management consultants and non-traditional engineering/consulting firms offer help with strategic planning.

- **Develop a go/no-go recommendation** and presentation for the policy board. Brief policymakers on business trends and environment, risk of status quo, challenges and opportunities, and vision of desired future. Request a decision and guidance on resources, schedule, and policy constraints.

"The strategic planning process needs to focus on the things that have to happen ***before you start evaluating options***...confirming the utility's vision and mission, assessing the policy board's position, assessing the utility's financial situation and the environment within which it operates."

Jon Eckenbach
JEA

Note: *This document is* ***not*** *intended to provide exhaustive how-to instructions for strategic planning. Instead, it explores water utility options for strategic planning, recommends a general framework, and presents results of the case study conducted with the Honolulu Board of Water Supply. The findings from this research will provide a foundation for follow-on strategic planning efforts the Research Foundation will conduct in 2003.*

Chapter 2: Opportunities and Constraints

Political environment

Public utilities in recent years have experienced a wholesale revision in the dynamics of the marketplaces in which they operate. Private-sector competition, bundled service offerings, and the deregulation of electric utilities have turned up the demand for rate containment – and have changed expectations about traditional utility services. At the same time, regulatory requirements, rehabilitation/replacement needs, system expansion demands, and limits on federal and state infrastructure funding have increased the pressure on utility revenues.

> "The strategic planning process is valuable in that it helps utilities look beyond the constraints they face. ***In their constraints lie opportunities.***"
>
> *Diana Gale*
> *Seattle Public Utilities*

To develop support at the political level for strategic planning, utilities must communicate its potential benefits to political leaders in their own language. The same pressures that drive up utility costs make it possible to realize many benefits from adopting new ways of providing service. Strategic thinking can open pathways to reduce operating costs, generate new sources of income, and prioritize investments based on what utility customers – *and voters* – feel is most important.

Strategic planning, in fact, presents a vital opportunity to demonstrate that utility goals and investments are aligned with the priorities of stakeholders – from policy boards, political leaders, and regulators to utility staff, customers, and the community at large. The journey toward alignment first requires a recognition that *traditional utility priorities do not represent all stakeholders.* For example, while a utility may prioritize investments in system redundancy and in improving treatment to exceed regulatory requirements, its stakeholders may be more supportive of investments in community amenities or system security – or they may prefer to accept the status quo to avoid rate impacts. As another example, graphic illustrations of how an asset management program can overcome service backlogs and control rates over the long term can build political support.

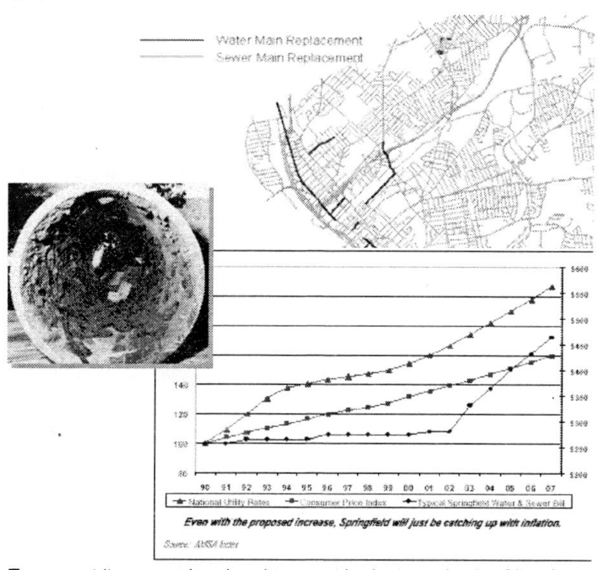

To support its renewal and replacement budget needs, the City of Springfield, Massachusetts, implemented an information campaign that linked system requirements to overarching community goals. The result was approval of a $70 million capital program that calls for annual rate increases of 11 percent over the next 5 years.

Involving stakeholders in setting priorities is not equivalent to ceding control over utility management and decision making. Instead, it is a means of building trust and smoothing the path for implementation of needed projects.

As with all stakeholder involvement, clearly defined boundaries for stakeholder participation help to manage expectations. Constraints and "sideboards" for strategic planning must be defined in two directions – both for stakeholder involvement and for the strategic planning process overall. For example, utilities must look to policy and community leaders for guidance on the range of acceptable rate impacts and the acceptability of potential directions such as regionalization or organizational restructuring. Because it is critical to pinpoint these sideboards at the outset of strategic planning, this activity is represented early in the planning process described herein.

The need for leadership commitment

Strategic planning can be a means of building utility management credibility – *if* the utility's policy board clearly supports the process. To this end, the policy board *must* be enlisted as a key sponsor of the strategic planning process at its outset.

"The leadership of the organization ***has to own the process.***"
Diana Gale
Seattle Public Utilities

To get policy board commitment, utilities must communicate clearly to policy boards what to expect from the strategic planning process in terms of goals, activities, schedule, and anticipated outcomes. The policy board's involvement throughout the process should be defined explicitly – and the utility should adhere to the planned activities and schedule. The strategic planning process described in Chapter 5 suggests key points for policy board involvement and commitment. Clear definition of the policy board's role and opportunities for review should limit iterations and start-overs, helping to preserve utility management credibility.

While costs and competition for local government funds continue to rise, securing funding for strategic investments may seem virtually impossible. ***Quite the opposite may be true***, if the strategic planning process links options clearly to definable community benefits.

Once expectations are defined, it is critical to maintain a "paper trail" throughout strategic planning, documenting policy board concurrence and commitment on decisions made *and* linking those decisions back to policy goals. This practice provides utility management a measure of protection against "democratic paralysis" – time-consuming revisiting and questioning of decisions.

In summary, meaningful and sustained commitment of leadership is required for strategic planning success. If the utility's policy board does not buy into the process and the decisions made along the way, the investment in strategic planning likely will be futile even if the decisions and supporting logic are sound.

Annual planning cycle and requirements

Just as a utility's goals must evolve in a constantly changing environment, strategic planning must be an iterative process. Considering the pace of change, a utility cannot develop a strategic plan that is immune to shifts in market and community dynamics. Therefore, the strategic plan must be revisited as part of – and in fact as a guide to – the annual planning cycle. For the same reasons, the strategic planning process must be simple, understandable, flexible, and *sustainable.* While a sustainable strategic *plan* implies an unrealistic degree of permanence, the *process* must be sustainable to ensure its continued viability.

The requirement for sustainability dictates that the strategic planning process be integrated smoothly with, rather than disrupting, the utility's existing annual planning processes. In fact, the strategic planning process optimally becomes an umbrella for all other planning processes, bridging the utility's goals to its annual budget.

While it typically focuses on multi-year objectives, strategic planning must flow down into the annual budgeting process to achieve a meaningful link between the utility's strategic goals and its day-to-day activities. The strategic planning process recommended in this document provides for a clear, recognizable link between the utility's agreed-upon mission, vision, and goals and the recommended investments, down to the individual operating unit level.

Importantly, the strategic planning process must reflect a realistic assessment of the utility's financial constraints and the degree of flexibility and risk that is politically acceptable and defensible. In some instances, utilities' "freedom to act" is defined in part by their compliance with prescribed limits on prospective rate increases. Other utilities may be able to secure near-term rate adjustments to make investments in system expansions, upgrades, and renewals and replacements. Still other utilities may have sufficient financial and political flexibility to pursue highly entrepreneurial investments. By integrating strategic planning processes with financial planning processes, utilities avoid violating prevailing financial constraints. Rather, in effect, existing financial policies are overlaid and new policies to support strategic direction are promulgated. In the changing utility marketplace, this linkage affords utilities the opportunity to more explicitly and consistently manage risks and balance emerging strategies to preserve financial health and integrity.

The strategic planning process must reflect a ***realistic assessment of the utility's financial constraints*** and the degree of flexibility and risk that is politically acceptable and defensible.

The level of effort and detail required for strategic planning may be defined by each utility's individual situation. For utilities addressing issues of significant risk and divergent stakeholder opinions, a more complex and defensible analytical approach is prescribed, alongside the requisite effort to develop strong supporting data. Utilities facing less complex decisions and opportunities can default to a simpler approach.

Similarly, the tools used to support strategic planning will vary depending on the level of risk and controversy involved. Strategic planning tools should "infiltrate" the overall planning process, with the goal of maintaining continual focus on strategic goals throughout all planning activities (master planning, capital planning, etc.).

"An effective strategic plan is a ***living document*** with a definable window - and it is revisited on a regular basis."
Walter Bishop
Contra Costa Water District

The recommended strategic planning process relies on a formal decision framework to catalogue stakeholder priorities. Using this decision framework to guide subsequent planning activities, the utility can justify its decisions by citing their linkage to stakeholders' stated priorities. This decision framework can make use of either intuitive information on the utility's opportunities, constraints, and performance metrics, or information developed through rigorous data collection and analysis. Among the tools provided herein to facilitate this process is a

Business Planning Template, presented in Step 5 of the strategic planning process (see Chapter 5 and Appendix B), which is to be completed by each of the utility's operating units as a precursor to capital planning.

Linkages to other agencies

Because strategic planning may involve consideration of non-traditional utility roles, and because utilities are influenced by a host of other entities, the strategic planning process cannot be accomplished in a vacuum. In addition to its policy board, the utility must define the appropriate level of integration and involvement for other agencies who influence, or are influenced by, the utility's activities. Regulatory agencies, local political and jurisdictional leadership, other utilities within the same or neighboring jurisdictions, and utilities with which services or products are exchanged are among those entities whose involvement will be required for strategic plan success.

A utility cannot realistically operate/make investment decisions as an independent entity if it is part of a local government organization. Rare is the public or private water utility that functions with complete independence from local or regional political and community leadership.

In addition to the obvious need for coordination among agencies, certain strategic investment options may heighten coordination requirements. For example, regionalization of utility facilities and functions is a strategic investment option that is increasingly common as cost control pressures persist. Other options, such as the production of bottled water or the provision of laboratory services on a contractual basis, requires close coordination and the maintenance of healthy working relationships with related agencies.

Clearly, the need for coordination with other agencies will vary significantly among individual utilities. The strategic planning process described in this document does not include recommended steps for inter-agency coordination beyond noting the need for it.

The following chapter explores options for stakeholder involvement in strategic planning. Utilities should consider inter-agency coordination a component of stakeholder involvement, but more specific coordination activities will be required for the investigation and prioritization of investment options for which implementation would require inter-agency cooperation and assent.

Chapter 3: Involving Stakeholders in Strategic Planning

At the heart of strategic planning is defining how a utility can better meet the needs of existing and future stakeholders. **The strategic planning process is *driven* by stakeholders' perceptions of value.** Strategies define how the utility will leverage market opportunities and address changing conditions to deliver valued services. Successful strategies are those that meet or exceed stakeholder expectations; therefore, it stands to reason that stakeholders must be involved in defining "value" and the means by which it is measured.

Considerations for strategic planning

Without involving stakeholders at critical junctures in the strategic planning process, utilities are developing candidate investment options in a vacuum. In so doing, utilities run the risk that ideas that initially seemed plausible are "dead on arrival" with employees, the policy board, and customers. Chances are a utility would rather avoid going back to the drawing board after investing time, energy, and resources in the strategic planning process. Utilities that develop strategic plans without obtaining some level of stakeholder buy-in along the way do so at their own peril.

"Planning for stakeholder involvement up-front is ***critical*** – and getting the policy board's buy-in to the process is key. ***Utilities must make an investment*** in getting their staff and policy board involved in the process."

Julius Ciaccia
Cleveland Division of Water

Ideally, utilities already have some form of ongoing public involvement in place, which can strengthen and feed into the strategic planning process. Many utilities have instituted standing Citizens' Advisory Committees, regular customer satisfaction surveys and focus groups, and involvement/information programs typically associated with major infrastructure projects. Utilities without such programs in place will have limited opportunities for obtaining feedback on the prospective appeal of investment options identified in strategic planning – which may translate into delays while the utility takes time to identify stakeholders, establish relationships, and solicit meaningful feedback. This was an issue for the Honolulu Board of Water Supply.

Different stages of strategic planning require different levels of stakeholder involvement. In general, involvement will be different from, and at a higher level than, the approach used for capital project implementation. Stakeholders at all levels should be involved in confirming the utility's vision, mission, and strategic goals, and there should be focused stakeholder involvement in developing the fundamental evaluation model (Step 3 of the recommended Strategic Planning Process, as described in Chapter 5).

The process of evaluating strategic investment options, however, will be (at least at first) primarily internal to the utility. "Blue sky" speculation, particularly when there are sensitive political ramifications, probably is best deliberated with a limited audience. As implementable ideas begin to develop, the utility can float trial balloons with key stakeholders. This is a delicate process, requiring sophistication and diplomacy. Done properly, it can mean a surer path to success.

Stakeholder identification

The first step toward defining the stakeholder values that will guide strategic planning is to determine who the stakeholders *are*. In so doing, utility managers must consider groups of individuals who are influenced by the utility's decisions, as well as individuals who have the potential to influence strategic plan implementation, inside and outside the utility, in both positive and negative ways.

Employees. In many ways, utility employees are the most critical stakeholder group of all. They can make or break the strategic planning process by embracing it and working to make it happen, by ignoring it and allowing it to die through inertia, or by actively obstructing it. Employees also serve as ambassadors to the community, considering the network of neighbors, friends, families and social groups to whom they carry the utility's messages.

Several different groups of employees should be treated as distinct stakeholders, including:

- Executive management
- Division managers
- Mid-level managers
- White collar workers
- Support and clerical staff
- Blue collar workers
- Union leadership

Vendors. Vendors, consultants, and contractors make up an often-overlooked group of stakeholders – those who do business with the utility. The success of their businesses is facilitated when the utility is running smoothly and sound decisions are being made in a timely fashion. It is worth scanning procurement department information to see which companies are doing business with the utility, and making sure those companies are on mailing lists for newsletters and other public information. Some of these businesspeople may be community leaders; it might be of benefit to know them well and solicit their ideas and feedback at appropriate points in strategic planning.

Policy board and community leaders. A utility's "authorizing environment" is the group of stakeholders from which the utility derives power and authority. The most immediate and obvious group is the policy board – those who are empowered by the community to vote on utility budgets, set rates, and establish the policies by which the utility is run. Other stakeholders are also part of the authorizing environment, depending on the form of government in the community, such as

- Mayor
- City council or selectmen
- County commissioners or supervisors
- Town meeting
- City/town manager
- Other municipal boards with oversight/approval power
- State auditor or inspector general
- State legislature
- State departments and regulators
- Federal regulators
- Congressional delegation

Depending on the level of detail the utility is taking with strategic planning, some of these individuals may be helpful as allies in developing and implementing a strategic plan.

Public. Unlike many other utility and government functions, water utilities provide services that touch everyone in the community; as a result, the external stakeholder group is unusually broad. To avoid being overwhelmed by the task of involving external stakeholders, it is helpful to categorize all possible stakeholders in terms of interests and potential effect on the utility's decision-making processes, and then to focus on those who would be most interested, and most likely to provide constructive feedback, for the strategic planning process. Categories of external stakeholders include

> "Customers and community are ***separate and distinct stakeholder groups***. They should be addressed individually in strategic planning, particularly in the context of utility goals and performance metrics."
>
> *John Huber*
> *Louisville Water Company*

- Economic (chambers of commerce, business groups, real estate developers, property owners and managers, industrial and commercial customers)
- Consumer (ratepayer/taxpayer groups, low income advocates)
- Civic/community (neighborhood associations, minority groups)
- Educational (academic institutions, teachers)
- Public health professionals (hospitals, medical and public health officials)
- Environmental advocates (land use groups, watershed associations)
- Media (print, radio, TV – commercial and cable, web-based)

As the utility proceeds in identifying stakeholders, it may help to develop "tiers" – a broader list of people who might be included on a mailing list, and a subset of people who have influence, internally and externally, to target for more direct involvement with the strategic planning process.

Outreach and involvement plan

Once stakeholders have been identified and the broad parameters of the stakeholder involvement program have been set, it is time to develop an outreach and involvement plan. If the utility already has a strategic communication plan, it is important to coordinate it with strategic plan outreach efforts to avoid confusing stakeholders with mixed messages and to leverage both efforts more effectively. If a utility does *not* have a formal communication plan, the strategic planning process can be a good vehicle for developing one.

Recommended activities in outreach and involvement for the strategic planning process include

1. Identify the internal and external stakeholders who may influence, or be influenced by, the outcomes of the strategic planning process.
2. Establish objectives for stakeholder outreach and involvement plan.
3. Review existing information from customer and public opinion surveys and other research.
4. Develop a focus group of key stakeholders from a cross section of the stakeholder types identified above; some utilities may have a "standing" focus group in place, but its membership will need to be reviewed in the context of strategic planning.
5. Develop a questionnaire for in-depth interviews with the focus group; schedule and conduct interviews, preferably with teams of two.
6. Synthesize input from focus group interviews to inform the development of investment options.
7. Hold small group meetings with internal and external stakeholders to confirm the organization's vision, mission, and goals, and to explain the strategic planning process.
8. Once investment options have been evaluated and prioritized, share the preliminary results of that evaluation with the focus group, presenting potential options in the form of trial balloons.
9. Once implementation plans have been drafted, invite review and comment from the focus group.
10. Close the loop by sharing the final strategic plan via an internal and external communication plan, explaining the process, stakeholder involvement, the outcome, and how stakeholder concerns were addressed in the final product.

Integrating stakeholder input

The following figure depicts suggested points for integrating stakeholder input into the recommended strategic planning process. More details are provided in the step-by-step strategic planning instructions included in Chapter 5.

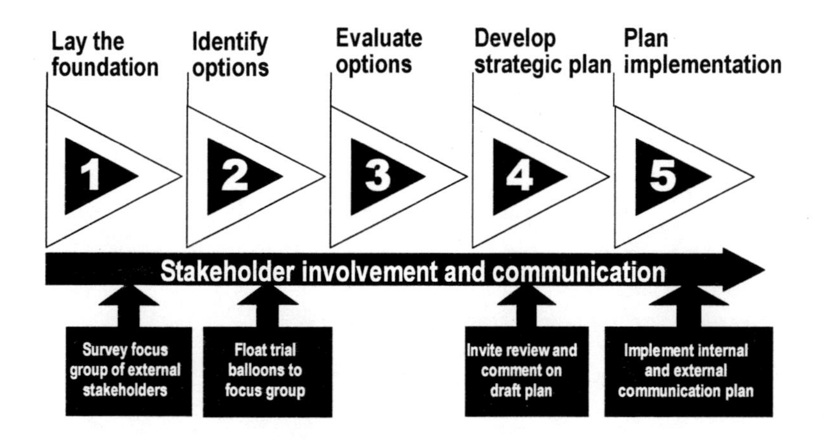

Stakeholder Input in Strategic Planning Process

Strategic plan communication

Once strategic planning has been completed, the outcomes may be documented for public consumption. It is recommended that utilities document strategic plans in an information piece using a simple, accessible format. This document should summarize the planning process and outcomes, while not necessarily documenting all the analyses used to define investment options and determine recommended strategies.

Each utility must consider its unique circumstances in determining how much information supporting the strategic planning process it is practical or advisable to publish. The presence of competitive interests may argue against public documentation of a utility's detailed competitive strategies. Conversely, publishing a sanitized strategic planning document that focuses on platitudes, without communicating a realistic vision and strategic direction, can put a utility's credibility at risk. Each utility must know what informational aspects of strategic planning may remain within the purview of senior management, considering applicable public meeting and right-to-know laws.

The printed information piece may be envisioned as a document placed on the coffee table in the utility's reception area, or perhaps handed out at a booth during Drinking Water Week, or presented to City Council. The plan should be concise and brief, as most audience members will be interested in but not necessarily obsessed with the details. Graphics should be used in place of words whenever possible, including photos of utility staff at work. Also recommended is a 10- to 15-minute presentation, describing the agency's vision, mission, goals, and highlights of the strategic plan.

Once the printed piece is developed, it can be posted on the utility's website (external and internal, if available) and mailed or e-mailed to the already assembled list of internal and external stakeholders, with a cover note from the utility's executive director. This is a good opportunity to invite commentary and feedback to help improve the process and the plan in the next year.

"Understandability is the key... there has to be another way for people to ***visualize the end result***, rather than having to read through an entire strategic planning document."

Adam Kramer
Minneapolis Public Works Department

Ideally, senior utility managers will take the presentation on the road, presenting it to groups of employees at staff meetings, to meetings of local business groups and civic organizations, to support a "speakers' bureau," and on a local access cable show, if one is available. Communicating the results of strategic planning is important to sustaining the effort. Utilities may need to develop separate, focused communication plans to support individual investments, particularly those that will represent a significant change in direction, entail a large capital outlay, or arouse a major divergence in stakeholder opinions.

Chapter 4: The Strategic Planning Process

The mandate for public utilities

As the institutional framework for water service delivery has moved from monopoly to market, water utility management increasingly has required an entrepreneurial perspective. More than ever, water utilities operate in a complex, competitive, and dynamic marketplace wherein survival is dependent on utilities' ability to protect and expand market share, manage risks, and improve customer satisfaction.

"Can we continue to run our business in the same way as we have in the past century? ***Not if we want to survive!***"
Cliff Jamile
Honolulu Board of Water Supply

In this context, water utilities must evaluate their market position, act strategically, and deploy resources to respond to rapidly changing market conditions. It is more important than ever for utility investments to yield expected returns, diversify risk, and protect market share. Undoubtedly, utilities unable to recognize and respond to the demands of market discipline will continue to be challenged.

In the 1990s, utility management focused largely on a singular aspect of the entrepreneurial challenge – responding to heightened competition. While this focus led to important advances and cost containment, it also brought the potential for compromising long-term strength and service capabilities in the name of short-term competitiveness.

"Strategic planning ***gives credibility*** to new investment options and management decisions."
Donna Kiyosaki
Honolulu Board of Water Supply

The true entrepreneurial challenge is not simply a matter of closing "competitiveness gaps." Rather, water utilities must develop long-term business strategies that reflect, and respond to, the changing marketplace – and afford opportunities for growth and development. This challenge is equivalent to that of any major, multi-functional business enterprise; at its core is the question of *how to invest limited resources to achieve highest returns with acceptable risks.*

Learning from private industry

More than ever before, changing market conditions suggest that utilities must align resource allocation decisions with organizational goals in business planning. With a tradition of monopolistic service delivery, many water utilities to date have not been challenged by the need to develop a *strategic* plan – and doing so may represent a significant shift.

As suggested by comparable utility services (such as telephone, electric, and cable), success is possible through efficient delivery of traditional services *in combination with* new or expanded service offerings. The strategic planning processes utilities use must acknowledge this dynamic context – suggesting the need to borrow techniques used successfully in the private sector. In particular, the fundamental principles of **portfolio management**, which has guided growth and sustainability of some of the most successful business enterprises[1] and investment houses, may provide important opportunities for water utility management.

Portfolio management is a business planning tool that considers the ***risks and rewards of strategic decisions*** in the context of the entire organization – and the marketplace in which it operates.

Portfolio management is an approach to business planning and risk management that considers the risks and rewards of strategic decisions in the context of the entire organization and the marketplace in which it operates. For water utilities, portfolio management provides an opportunity to integrate various planning functions under a single framework – one that directly links strategic business goals to the annual planning and budgeting process.

Portfolio management provides a unifying decision framework because it is designed specifically to consider investments with multiple attributes, risks, and requirements. Though water utilities face added complexity in that "returns" must be defined in both monetary and non-monetary terms, all major utility functions may be evaluated within this context. Many private sector utilities are already using portfolio management practices to address business challenges very similar to those facing public utilities.

In a portfolio management approach, utility managers evaluate each major utility function in terms of investment opportunities – and consider resource allocation decisions in the context of the utility's entire set, or portfolio, of assets. Portfolio management lets utilities balance the investment of public resources across a broad range of assets – hard capital assets as well as soft knowledge and customer service assets. In this way, utility managers can make informed business planning decisions in light of the total range of assets, considering their complex inter-relationships, rather than making decisions as part of discrete and often loosely connected planning activities. While private-sector portfolio management measures risk versus return, the public sector must expand on this relationship to measure how effectively different investment mixes, or "portfolios," meet utility and stakeholder goals (monetary and non-monetary).

Portfolio management affords particular advantages for water utilities because of the greater complexity of resource investment decisions. Whereas commercial interests make investment decisions by assessing potential monetary risks and returns, water utilities face the challenge of securing monetary and non-monetary benefits where risks are often more profound (e.g., public health, environmental degradation) and prospective returns are harder to measure. Water utilities must balance resource investments across multiple, competing priorities (e.g., environmental stewardship, support of economic development) to serve diverse stakeholder interests (e.g., development community, ratepayers, environmental interests, regulatory agencies). In these circumstances,

[1] For example, Anheuser-Busch, Chase, Coca-Cola, GM, and the Southern Company, to name a few, are characterized by extensive, and actively managed, investment and new product portfolios.

the use of portfolio management principles becomes even more compelling and will enable water utilities to

- Optimize asset use and investment opportunities
- Fully leverage resources to enhance competitive success
- Effectively manage risks and conduct risk-based decision making
- Effectively analyze monetary and non-monetary investment decisions within a single analytical framework
- Explicitly recognize benefits of asset allocation and risk diversification

The value of the AwwaRF-BWS tailored collaboration project is to demonstrate the use of portfolio management principles in developing a strategic plan, outline the complications and benefits of this planning perspective, and illustrate how *strategic* planning may be integrated into annual water utility planning processes, which today often do not link strategic goals to tactical resource allocations. In so doing, the tailored collaboration project offers insights into how utility managers may more effectively anticipate and respond to changing water utility management challenges and market conditions.

The basic steps of portfolio management are straightforward:

1. Identify the investment options available and assess their potential benefits, risks, and resource requirements;
2. Define benefits and risks in terms of their relationship to organizational goals;
3. Prioritize investment options within and across functional areas; and
4. Select a "portfolio" of investment options for inclusion in the utility's strategic plan.

Basic steps of portfolio management:

1. Identify investment options.
2. Define benefits and risks.
3. Prioritize options within and across functional areas.
4. Select a "portfolio" of investment options for inclusion in strategic plan.

The recommended strategic planning process for water utilities, introduced below, incorporates these steps into a flexible framework applicable to the public and private utility industry.

Recommended strategic planning process

Specific steps of the strategic planning process developed for this project (shown on following page) require utility management to assess the utility's market position, identify strategic investment options, prioritize investments based on the extent to which strategic goals are advanced, and establish a portfolio of options for implementation. The strategic planning process focuses specifically on those initiatives and investments that will *effect a change in the status quo* and allow the utility to *advance its market position* and meet any other strategic objectives. This process uses well-established evaluation techniques to quantify monetary and non-monetary impacts, manage stakeholder involvement, and prioritize investment options.[2]

Step-by-step instructions for the recommended strategic planning process are provided in Chapter 5.

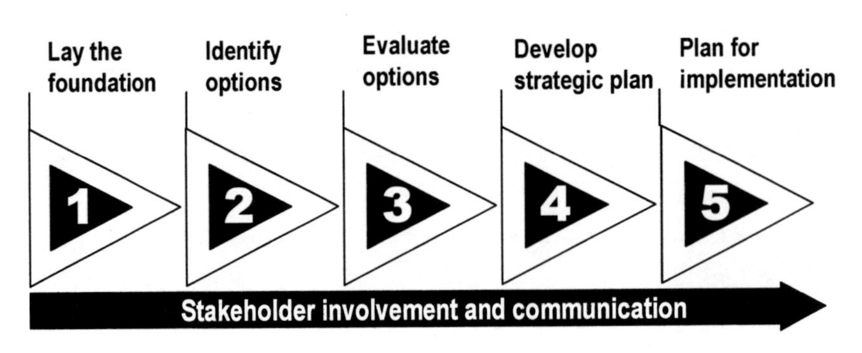

Recommended Strategic Planning Process

Integrating strategic planning into the existing planning cycle

For many utilities today, the connection between strategic goal-setting and budget development is tenuous at best. Annual planning processes often suffer from a disconnect between policy goals and the detailed tactical realities of line-item budgeting. A ***strategic*** planning process can establish the missing linkage by creating an "umbrella" underneath which to organize and align other utility planning activities. Accordingly, the AwwaRF-BWS strategic planning process is intended to better align, rather than duplicate, a utility's existing planning activities – and to provide a bridge between policy-level goals and the utility's annual budget development process. This process translates utility-wide strategic goals into specific tactical plans and performance measures for which individual utility operating divisions are accountable.

Because the strategic planning process is integrated into the overall utility planning cycle, and because it requires concise planning submittals that support budget development, the process does not duplicate but rather *supports* budget development efforts. The strategic phase of the planning cycle occurs before the tactical processes of budget development and organizational restructuring.

To be an effective instrument for resource allocation decisions, strategic planning must become a sustained process that sets a foundation, revisited annually, for other planning functions. As shown in the graphic that follows, the strategic planning process maps to the typical utility business planning process, affording a mechanism to move from goal setting to line-item budgeting.

Integrating *strategic* planning into the annual planning cycle is perhaps the single most important factor for ensuring long-term sustainability of a strategic plan. Products of strategic planning offer concise utility management tools to ensure that strategic goals are advanced and necessary resource allocations prioritized. In particular, the strategic planning process considers a planning phase in which utility

---

2 For a more complete description of each step of the strategic planning process, as well as more information on strategic investment portfolio construction see Appendix C.

management defines, in concise and measurable terms, the near-term and 5-year work program it will conduct to achieve (or advance toward) the utility's stated strategic goals. This work program should reflect a rigorous prioritization of strategic investment options – conducted in advance of budget submittals. It also should serve as a management tool to monitor strategically important initiatives, and should be readily updated or revised to effect responses to changing market conditions or organizational objectives.[3]

This strategic business planning process thereby addresses several opportunities for improvement of annual planning cycles that prevail in many utilities. The process

- Ensures alignment of strategic goals and objectives with divisional plans and budgets;
- Effectively prioritizes investment options; and
- Translates strategic goals to specific work assignments with defined performance measures and scheduled milestones.

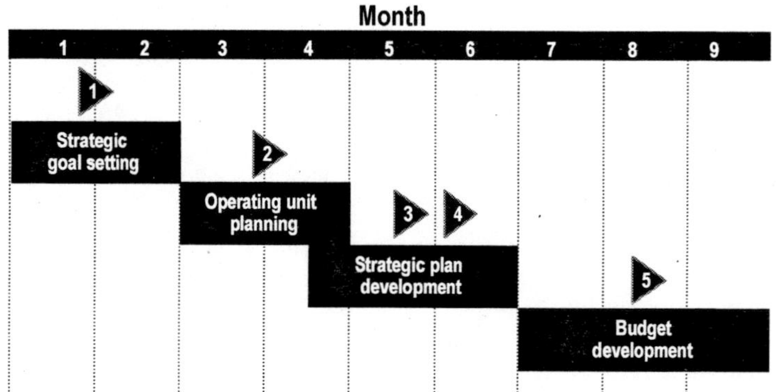

The steps in the strategic planning process (indicated by triangular numbers) are aligned with activities in the Honolulu BWS's annual planning and budgeting cycle (indicated by horizontal bars).

Participants in strategic planning

At some level, each of a utility's internal and external stakeholder groups ideally will have an opportunity to provide input into strategic planning. It is recommended that strategic planning be led by executive-level utility management, with precise points identified for policy board review and input, and specific deliverables required from members of the utility management team representing individual departments or operating units. Recommendations for providing other stakeholders opportunities for input are discussed in Chapter 3. In general, other stakeholders' input will have the greatest influence during Step 1 of the process, when strategic goals are confirmed and their relative weights determined. Stakeholder communication will continue to some degree throughout the planning process – culminating in distribution of a printed piece summarizing the adopted strategic plan.

In each step of the strategic planning process, a series of activities must be completed and decisions made within the utility – at the policy level, management level, and staff level. The instructions included in Chapter 5 are focused on recommended activities and decisions for each step, at each organizational level.

---

[3] Specific examples of templates used to develop components of the BWS strategic plan are provided as Appendix B. These individual operating unit submittals may be readily revisited and updated; milestones and performance measures related to each operating unit will be summarized in an enterprise-wide schedule of strategic initiative milestones.

Chapter 5:
Step-by-Step Guide to Strategic Planning

Utilities seeking to integrate strategic planning into their annual planning cycles can do so through a well defined, step-by-step process (illustrated below). While each of these five steps can be taken with different degrees of rigor and levels of effort, the process helps ensure that utility investments yield the greatest returns in terms of monetary and non-monetary benefits. This process also invites continuous improvement. In each planning cycle, utilities can build on and improve the information and processes used to complete steps in prior years.

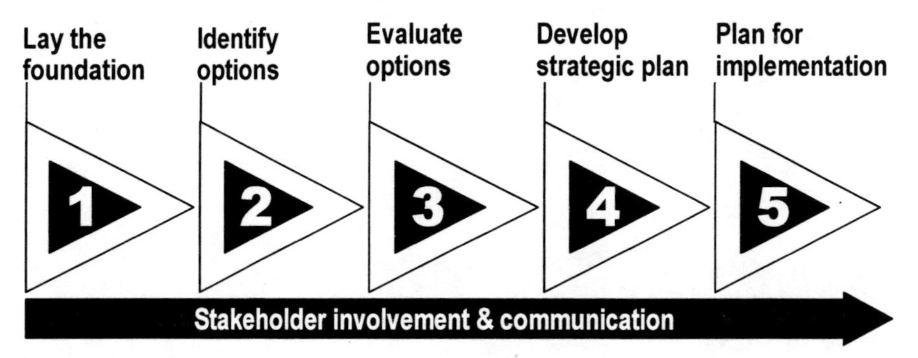

"Strategic planning is an ***evolutionary versus a revolutionary process.*** It requires achievable steps that people can adapt to within a controlled environment, without creating anarchy."

John Huber
Louisville Water Company

The basic purpose of each step is as follows:

1 – Lay the foundation. A solid foundation is essential for a viable strategic plan. Considering the political realities within which utilities operate, this step involves working with policy boards to define baseline conditions, policy directions, and boundaries. As strategic planning is integrated into the annual planning cycle, this step will involve revisiting policy directions based on changes in conditions, policy board members and priorities, and stakeholder expectations.

2 – Identify options. Working with stakeholders, the utility assesses community priorities and confirms its vision, mission, and goals in light of Step 1. Strategic opportunities begin to take shape in this step.

3 – Evaluate options. Resource investment strategy options are clearly defined, an evaluation model constructed that reflects stakeholder priorities, and options evaluated and scored.

4 – Develop strategic plan. Ranked strategic investment options are grouped into "investment portfolios" and prioritized to form a complete strategic plan, recognizing synergies across operating units.

5 – Plan for implementation. Implementation and contingency planning is conducted to help assure the strategic plan is interwoven with the utility's business processes. Final approval is solicited from the policy board, and implementation begins. A system for monitoring progress is put in place to facilitate annual updates of the strategic plan.

In each step of the process, a series of activities must be completed and decisions made based on their outcomes. As decisions are made in each step, those decisions provide a foundation for subsequent steps. Documentation of those decisions helps ensure that the final strategic plan is defensible and can be implemented.

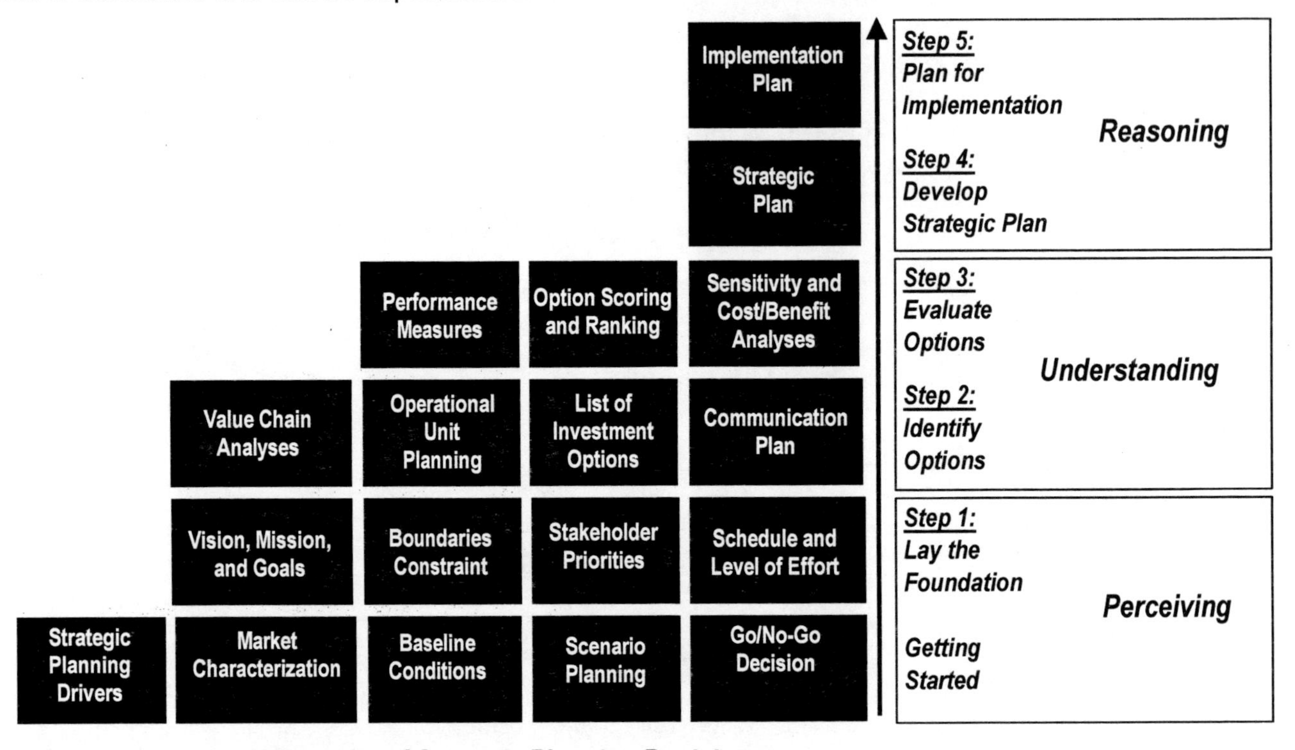

Hierarchy of Strategic Planning Decisions

On the following pages is a summary matrix of the key decisions, activities, tools, and outcomes of each step in the strategic planning process, including the preliminary activities discussed in Chapter 1: Introduction. Following this matrix is a detailed description of the purpose of each step, and of the case study experience in strategic planning with the Honolulu Board of Water Supply.

Note: *This document is* ***not*** *intended to provide exhaustive how-to instructions for strategic planning. Instead, it explores water utility options for strategic planning, recommends a general framework, and presents results of the case study conducted with the Honolulu Board of Water Supply. The findings from this research will provide a foundation for follow-on strategic planning efforts the Research Foundation will conduct in 2003.*

Strategic Planning Process Overview

Step		1	2	3	4	5
	Getting Started	Lay the Foundation	Identify Options	Evaluate Options	Develop Strategic Plan	Plan for Implementation
Objectives	Build business case for strategic planning. Obtain go/no-go decision from policy board.	Confirm utility vision, mission, and goals. Assess stakeholder priorities (internal/external).	Identify viable strategic investment opportunities for utility's investment portfolio.	Evaluate and rank strategic investment options.	Group and prioritize strategies into a completed strategic plan, recognizing synergies across options and operating units. Obtain go/no-go decision from policy board.	Develop implementation and contingency plans. Charter teams for implementation. Communicate with internal/external stakeholders.
Key Actions & Decisions						
Policy Board	Provide preliminary buy-in for development of strategic plan. Provide go/no-go decision and guidance on resources, schedule, constraints.	Confirm utility vision, mission, and goals. Reach consensus on policy boundaries for strategic planning. Review evaluation criteria.	Consider the range of strategic investment opportunities and assess risks associated with change versus status quo; provide feedback and direction.	Review evaluation results and provide feedback.	Review draft strategic plan and provide feedback.	Sponsor implementation and communication of strategic plan.
Senior Mgmt.	Consider key questions, policy issues that influence strategic planning success. Assess resource requirements, financial/managerial capability to complete strategic plan. Conduct scenario planning; assess business risk of status quo. Brief policymakers on business trends, environment, players. Develop go/no-go recommendation for policy board.	Revisit and confirm utility vision, mission, and goals. Assess competitiveness of compensation/benefits, capacity to attract/retain required expertise. Summarize board policies to set boundaries; identify needed change. Define evaluation criteria and decision boundaries.	Review input from operating units; define the range of potentially viable strategic investment opportunities.	Score and rank strategic investment opportunities. Conduct sensitivity analysis and cost/benefit analysis. Define recommended actions. Develop presentation for policy board and focus group.	Bundle combinations of investment options. Complete draft strategic plan and present to policy board; finalize plan. Establish milestones and reporting mechanisms for the organization based on recommended strategy.	Develop action plans for implementation – rate/financing analysis, staffing adequacy.

(continued)

Step		1	2	3	4	5
	Getting Started	Lay the Foundation	Identify Options	Evaluate Options	Develop Strategic Plan	Plan for Implementation
Mgmt. Team	Review definition of a strategic plan. Define baseline conditions, issues affecting operational success. Describe current business settings: constraints and opportunities, within and beyond utility's control. Estimate schedule and level of effort to prepare strategic plan.	Evaluate status of operational planning processes. Assess operational strengths and weaknesses in context of strategic investment opportunities/threats. Identify stakeholder groups; define their role in strategic planning. Develop communication plan (internal and external). Develop stakeholder focus group; survey to assess priorities.	Identify strategic investment options by operating unit, aligned with utility goals, and associated performance metrics (see template in Appendix B). Conduct value chain analysis. Float trial balloons to focus group; adjust options list if perceptions are negative.	Develop performance measures and evaluation model. Score and rank strategic investment opportunities.	Identify and quantify operational adjustments and systems/data needed to measure progress in implementing strategy.	Develop detailed implementation schedule for final strategic plan, by operating unit. Implement internal and external communication plan.
Public/ Stake-Holders	Solicit input from unions/employees.	Survey focus group of customers, suppliers, regulators, and community representatives to assess stakeholder priorities and preferences, desired service levels, willingness to pay for enhanced services.	Obtain focus group feedback on trial balloons. Solicit union/employee input on vision/mission/goals.	Obtain focus group feedback on recommended actions.	Obtain focus group feedback on draft summary of strategic business plan.	Implement internal and external communication plan.
Tools/ Techniques	Scenario planning List of questions to consider (see Chapter 1) Industry business journals and reports	Strengths/Weaknesses/ Opportunities/Threats (SWOT) analysis Stakeholder survey Communication plan	Business planning templates (see Appendix B)	Decision analysis tools for scoring, ranking, sensitivity and cost/benefit analyses	Rate/financing analysis Mechanisms for monitoring performance	Implementation scheduling and tracking tools

(continued)

Step		1	2	3	4	5
	Getting Started	**Lay the Foundation**	**Identify Options**	**Evaluate Options**	**Develop Strategic Plan**	**Plan for Implementation**
Outcomes	Go/no-go decision Policy direction and definition of boundaries	Vision/mission/goals and evaluation criteria Organizational strengths, weaknesses, opportunities, threats Initial performance measures Stakeholder survey results (internal/external) List of business opportunities	Business planning templates by organizational unit Communication plan (internal/ external) and supporting collateral materials Focus group feedback on trial balloons	Performance measures and evaluation model Ranked business options	Draft and final strategic plan Action plans for implementation by operating unit Focus group feedback Go/no-go decision from policy board	Budgets Staffing Workplans by task Schedule/milestones Monitoring/reporting protocol
Findings	It often is difficult to assess risks of maintaining status quo, due in part to myths about the direction in which the market is moving. Utilities must address cultural resistance to change at all staff levels. Managers need education on business planning theory to develop an understanding of the need, and potential, for change. Existing data often are inadequate to define baseline conditions. It is critical to integrate strategic planning with the annual planning cycle.		Utilities need to develop business opportunities *exhaustively and creatively*, challenging traditional thought.	It is difficult to define measurable performance criteria with limited available data; this suggests a need for initial strategic investments in enhanced data collection. Middle management needs guidance on integrating strategic opportunities into their business plans. There is a variable appetite/perceived need for evaluation analytics. Water utilities have an inherent resistance to semantics of private industry.	*Facilitated* portfolio management is an easily applied technique that can be of great value to water utilities.	Middle management tends to prioritize day-to-day tasks/crises over strategic activities. Middle management often exhibits hesitation to lead, due in part to fear of blame or "rocking the boat."

(continued)

Step		1	2	3	4	5
	Getting Started	Lay the Foundation	Identify Options	Evaluate Options	Develop Strategic Plan	Plan for Implementation
Future Research	Guidance on determining market directions, risk of status quo Research on the future of the industry, what role competition plays		Assessing changing expectations of customers and decision makers Survival in a radically changing marketplace: Lessons learned from electric utility, telecom, other industries Beyond benchmarking: Appropriate performance measures for a changing industry	Evaluation, quantification and management of risk (for both traditional and non-traditional strategic investment opportunities) Beta testing of entire strategic planning process	Strategic planning in other industries Communication of strategic plan results Full watershed planning – implications	Tools for feedback on progress of strategic plan Survey to assess how many utilities perform strategic planning and update the plan regularly Techniques for fostering strategic thinking at all organizational levels Definition of clear linkages between long-term strategic planning and annual budgeting processes

Step 1 – Lay the foundation

Purpose of this step

To define boundary conditions for strategic planning; confirm the utility's vision, mission, and goals; and assess stakeholder priorities to guide strategic investment decisions.

Description

Step 1 provides the foundation for the strategic planning process. It involves establishing or confirming the utility's vision and mission, and the fundamental goals to which the utility's discretionary resources will be directed. This requires a careful assessment of the utility's political and stakeholder environment, because the utility's fundamental goals will become the evaluation criteria that define how the utility addresses and prioritizes stakeholder concerns. This step also is intended to invoke a strong commitment of utility leadership and policy-level decision makers.

Defining the utility's fundamental goals, and thereby the criteria for evaluating investment options, provides the context for annual planning processes. As the majority of strategic planning is conducted at the front end of annual planning efforts (as noted in Chapter 2), defining what the utility's allocation of resources is *intended to achieve* offers guidance for planning efforts. For example, highlighting the link between planned investments and achievement of utility goals will make a more compelling argument for budget justifications than calls for preservation of the status quo. Likewise, stakeholders are more likely to support new initiatives if they are shown to be consistent with the utility's stated goals.

Also in this step, the utility must consider how it will physically integrate the strategic planning process with other ongoing planning initiatives:

- Capital improvement plan
- Future water supply plan
- Financial plan
- Facilities master plans
- Asset management plan
- Regulatory compliance plan
- Others

Providing mechanisms for transferring information among planning functions is of critical importance.

Key activities for Step 1

The diagram below includes the activities recommended for completion of Step 1. This diagram is organized to display the relative sequence of events, from left to right. From top to bottom, the activities are grouped by organizational level: policy board activities and interaction are shown at the top, with senior management, the utility's management team, and stakeholder activities shown beneath. It is generally expected that the activities in Step 1 will take approximately 1 month to complete. A description of Step 1 completion for the Honolulu BWS is included on the following pages.

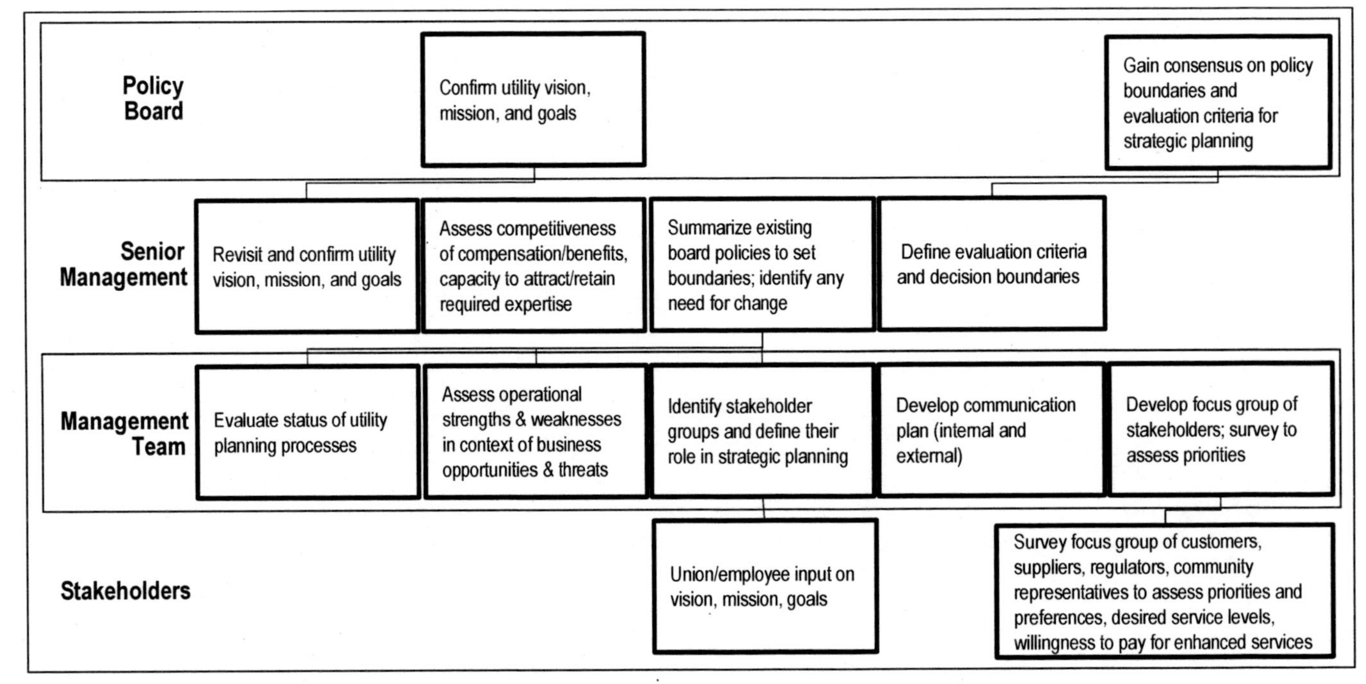

Further details on recommended steps for strategic planning will be developed in a follow-on Research Foundation study.

Applicability to water utilities

The increasingly competitive and resource-constrained environment within which utilities operate, as well as heightened performance expectations of stakeholder groups (customers, regulators, environmental advocates), necessitate a clear delineation of strategic goals.

Goals and objectives provide a compelling statement of the utility's purpose and direction, offering both internal and external stakeholders a sense of vision and direction. They may be an effective instrument in communicating the imperatives of changing market conditions.

Case study description

As part of this project, the team conducted a step-by-step strategic planning process with the Honolulu Board of Water Supply. This process was documented as a case study to ground-truth prescribed tools and processes. The BWS is publishing the resulting strategic plan for dissemination to key stakeholders; a draft version of the plan is included as Appendix A to this document.

In this section, for each process step we describe our findings and experience related to the conduct of the Honolulu BWS strategic planning case study.

Findings

- Policy goals and objectives may be developed without consideration of prospective use in the evaluation of strategic investments. Policy goal statements are often imprecise, interdependent, and designed for cosmetic appeal rather than analytical rigor.
- The relative importance of utility fundamental goals often is not addressed explicitly, complicating subsequent weightings required in strategic investment evaluation.
- Differences between mission, vision, values, goals, and objectives are not commonly understood among utility management personnel.
- Unavailability of benchmark comparison data on utility performance may complicate the definition of specific objectives.

Experience

Because the Honolulu BWS was in the midst of an ongoing restructuring and quality improvement program, its definition of *resource portfolio goals* relied heavily on the initial outcomes of this organizational development process. Utility leadership referenced revised vision, mission, and goals statements, developed through this process, to aid in the definition of resource portfolio goals. Using these statements as a starting point, further definition of fundamental objectives associated with each resource portfolio goal were identified. These are illustrated, in abbreviated format, through the construction of a "value hierarchy."

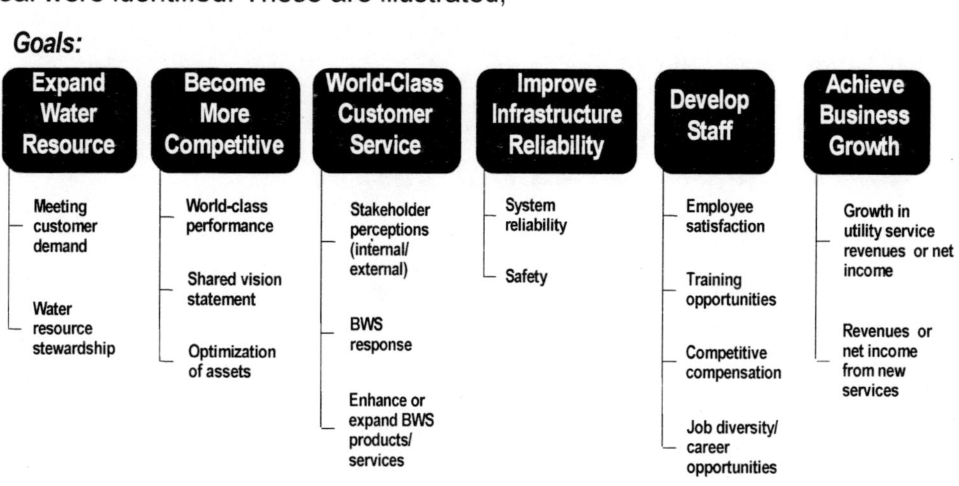

Honolulu Board of Water Supply Value Hierarchy

The process of determining fundamental objectives was relatively straight-forward. For each of the utility's lofty stated goals, several questions were posed including:

- What does this goal mean in practical terms?
- How would utility stakeholders define this goal?

- What is the relative importance of each of these goals?
- How will these goals direct the utility's allocation of resources?

These questions led to a variety of revisions and consolidation in the listing of fundamental objectives. Ultimately, the resource portfolio goals and associated fundamental objectives provided a workable foundation for evaluating strategic investments.

The BWS's value hierarchy highlights a number of common characteristics of typical utility goal statements. These include reliance on terms and concepts that, while popular, have broadly defined and imprecise connotations. Phrases such as "world-class performance" and "competitiveness" convey important commitments to excellence. Listing fundamental objectives, and establishing performance measures (in Step 3), enabled more specific meanings to be defined.

Perhaps more importantly, the BWS's goal statements – constructed in advance of entering into the strategic planning project – largely reflected the interdependencies of utility management challenges. For example, it is arguable that "becoming competitive" is a prerequisite to "world-class customer service."

Some basic rules for development of resource portfolio goals include:

- **Comprehensive**. The utility's goals should reflect the entire scope of its endeavors, from effective management of existing assets to new initiatives aimed at enhancing the utility's position in the marketplace.
- **Fundamental.** The utility's goals should reflect the organization's purpose – addressing imperatives for protecting public health and stewarding environmental, financial, and community resources.
- **Non-redundant and independent**. Redundant or interdependent goals complicate decision making. Determining how to invest resources requires evaluating the inherent tradeoffs. If goals are structured such that accomplishment of one goal is largely a consequence of accomplishing another, these tradeoffs are muted.

Step 2 – Identify investment options

Purpose of this step

To identify viable business opportunities for utility's investment portfolio.

Description

Once the utility has established (or confirmed) its fundamental goals and objectives, it is in a position to consider a broad spectrum of possibilities for investing its limited resources. In the increasingly competitive, and

Strategic investments may be identified from ongoing planning efforts, or through stakeholder initiatives.

integrated utility marketplace, this spectrum often does not resemble the resource investment choices faced by utility managers just a generation removed. Forays into bottled water, reclaimed water, energy generation, and even telecommunications infrastructure are a bellwether of changing times. Similarly, heightened concerns related to infrastructure security, asset management, and total watershed management impose new investment imperatives. Effective strategic planning requires that utility managers understand the breadth of the possibilities available to them both in the abstract and, more importantly, within their own utility context.

Step 2 is intended to help utilities define the range of investment options that may be appropriate for them. For some utilities, these options may be limited to expansion or enhancement of "core" utility functions, while other utilities may reflect a more entrepreneurial focus. Because utilities already are engaged in a multitude of individual planning efforts, one purpose of this step is to cast the outcomes of these planning processes in terms of strategic investment options. Another aspect of this step is to provide opportunities for staff and interested stakeholders to bring out-of-the-box thinking to the table. Ultimately, this step helps identify the set of options that the utility, its policy board, and its key stakeholders may support and view as consistent with the utility's mission and goals.

Key activities for Step 2

The diagram below includes the activities recommended for completion of Step 2. This diagram is organized to display the relative sequence of events, from left to right. From top to bottom, the activities are organized by organizational level: policy board activities and interaction are shown at the top, with senior management, utility management team, and stakeholder activities shown beneath. It is generally expected that the activities in Step 2 will take approximately 1 month to complete, but time will vary significantly based on the utility's chosen level of analytical

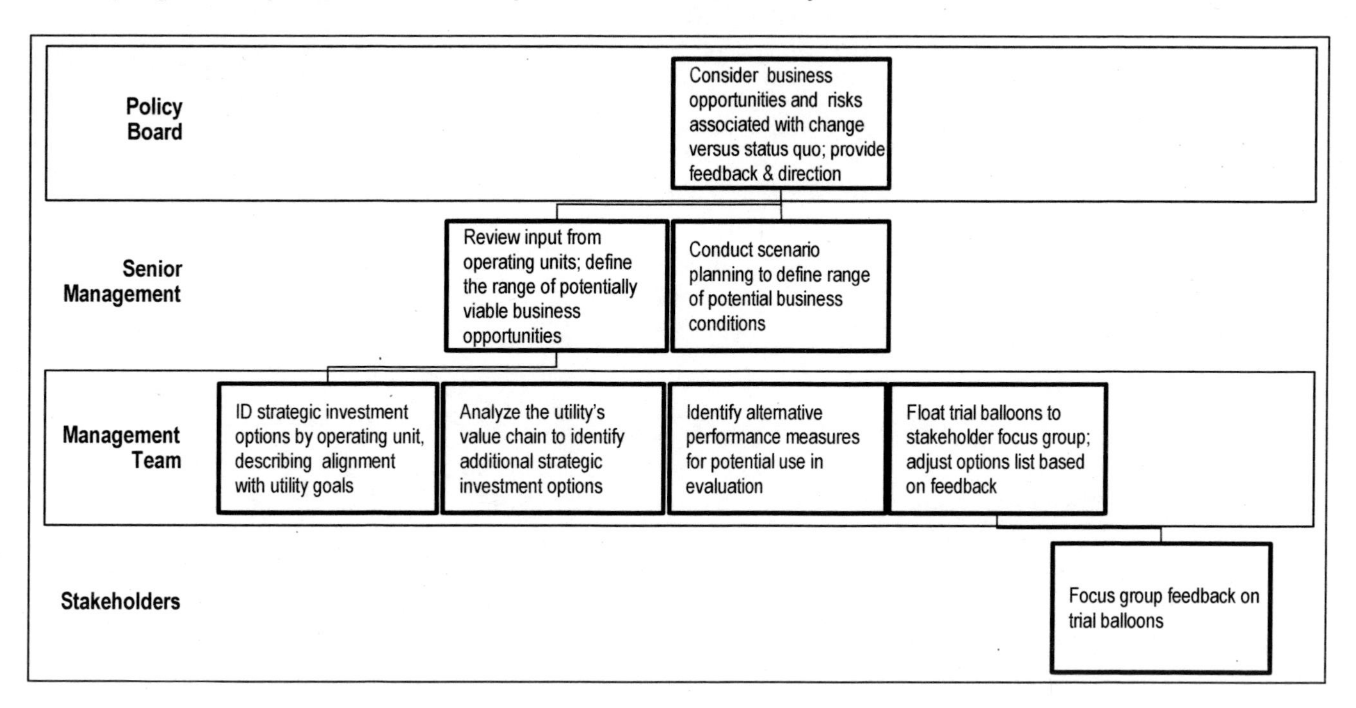

detail. A description of Step 2 completion for the Honolulu BWS is included on the following pages. Further details on recommended steps for strategic planning will be developed in a follow-on Research Foundation study.

Applicability to water utilities

It is important to consider the notion of a utility's resource allocations as "investments." Utility managers face a dizzying array of choices – to acquire neighbor systems versus renew and replace existing assets, to enhance information management systems or develop supply reserves. As identified in *Dawn of the Replacement Era: Reinvesting in Drinking Water Infrastructure* (AWWA, May 2001), and as evidenced in the push for total watershed management, utilities must consider these choices and meet customer demands in a context of shrinking budgets and failing infrastructure.

Viewing resource allocations as *investments that provide a measurable return* can help utilities survive in a competitive marketplace. If the utility identifies only a limited set of traditional investments, it may lose opportunities to enhance competitiveness and fulfill its potential. If it does not recognize the constraints within which it operates, it risks losing focus and credibility.

Strategic investments proliferate throughout the industry in examples such as regionalization (Louisville Water Company's system acquisitions), ventures into new services (Honolulu BWS's entry into the water reclamation market), and system revitalization (Orlando Utilities Commission's investments in system-wide automation).

Case study description

Findings

- Investment options largely reflect the vision and risk appetite of utility leadership. The BWS's transformation has been a consequence of a revitalized managerial style and commitment of BWS leadership.
- Most investment options were identified in the utility's existing planning processes, though institutionalization of a Business Development unit served to catalyze identification of strategic investment options.
- Data on the cost or potential performance of investment options typically was limited, particularly for strategic investments that involved expansion of products and services.

Experience

CH2M HILL compiled a set of strategic investment options for the BWS through a combination of information sources. These included interviews with BWS leadership and stakeholders, reference to the utility's budget proposal and planning studies, and documentation of the BWS's reengineering program (referred to as QUEST). This listing of strategic investments was confirmed by BWS leadership in a subsequent workshop discussion.

Characteristics of a *strategic* investment

- Creates a new customer service that enhances value to the customer, or
- Fundamentally changes the way an existing service is delivered to enhance value to the customer, or
- Provides a mechanism for managing monetary and non-monetary risks.

BWS Strategic Investment Options 2001

- Brand imaging / communications
- Customer billing & service offering
- Laboratory services
- Pump station retrofit/SCADA
- Fund RFIP (level A, B, C)
- Engineering/CM/Program Mgmt.
- Watershed protection partnerships
- Water resource development
- Desalination
- Ewa Shaft, other groundwater resource development
- Conservation/demand management
- Beretania/other real estate site developments
- System acquisitions
- Bottled water program
- Property management
- BWS site leasing
- Nu'uanu Reservoir recreation
- Financial asset management
- Investments
- Debt management
- Enterprise system - IT (FAS, CAS, MMS, GIS)
- Learning Academy

Finally, **scenario planning** was conducted in a facilitated workshop involving the utility's management team. In this workshop, the team defined expected market conditions under three potential scenarios: best case, worst case, and status quo. Within each of these scenarios, economic, political, social, and environmental factors were used to characterize market conditions. The resulting sets of market conditions provided a context for evaluating the robustness of strategic investment ranking conducted in Step 3.

Considerable time and effort was dedicated to distinguishing strategic investments from the daily resource allocations required to maintain utility service levels. **Strategic investments** are those allocations of resources that achieve one or more of the following criteria:

- Creates a new customer service that enhances value to the customer
- Fundamentally changes the way a utility delivers a customer service, thereby enhancing value to the customer
- Provides a mechanism for managing a utility's monetary and non-monetary risks

Perhaps most importantly, strategic investments are projected to result in measurable progress toward achievement of a utility's strategic goals.

As is characteristic of many utility organization, BWS's strategic investment options reflect the vision of the utility's top executives and the political environment within which they operate. BWS has enjoyed a relative transformation in recent years, now embracing the challenges of market competition. Options such as investment in reclaimed water systems are now being pursued, whereas previous leadership had considered such options beyond the utility's purview.

Notes on Scenario Planning

Scenario planning characterizes the business environment within which a utility will most likely operate over time and potential favorable or unfavorable changes that may affect business performance. Several scenarios may be developed to represent best case, worst case, and status quo conditions.

This planning technique may help determine strategic goals, articulate a proactive strategic direction, and define risk-management methods that will buffer the organization against adverse business environment changes. Indicators or "triggers" that forewarn the utility of a change in the business environment may be identified so that steps can be taken to adjust the adopted strategy rather than reacting after the fact.

In identifying strategic investment options, it was evident that in many cases little more than anecdotal information about the investment options of interest was available. In particular, research will be required to develop cost estimates for many of the most conceptually promising strategic investments. In addition, the team encountered some complications in identifying and characterizing investments as a consequence of uncertainties associated with the BWS's ongoing reengineering program. Among its other impacts, this program will enhance the BWS's ability to leverage selected market opportunities. However, conversions to new systems and business processes were in their infancy during the strategic planning process; as a result, BWS management and staff were uncertain as to what additional services may be rendered once new systems and processes are in place.

Similarly, the absence of more institutionalized stakeholder involvement limited the spectrum of options considered to be viable. Interviews with selected stakeholders and Board members suggested that support may be garnered for an even more expanded BWS role in a number of enterprises on the islands. Enhanced community communication and involvement will help strengthen community perceptions that the BWS is a logical choice for delivering these new services. It will also help establish an environment by which the BWS's stakeholders are chartered with providing input on candidate investment options.

Step 3: Evaluate investment options

Purpose of this step

To score, rank, and evaluate business strategy options for inclusion in the strategic plan.

"In the water industry, we tend to practice risk avoidance rather than risk management. We need to examine risk carefully, especially when there are opportunities involved, and not always avoid them."

John Huber
Louisville Water Company

Description

Step 3 involves building a model for evaluating the investment options, which is then used to rank and score the options. Of utmost importance is that this evaluation basis be aligned with the goals established in Step 1. Otherwise, the utility's investments may compromise rather than advance its market position.

Performance measures are established to gauge contributions of utility investments to strategic goals in discrete, measurable terms. Characteristics of sound performance measures are precision, scalability, and substantive content. Precise performance measures support the development of performance scales that allow clear, unambiguous "scoring" of strategic investments. That is, utility staff and external stakeholders alike generally will assign similar performance scores for given investments because the determinants of scoring are so clearly defined. Scalability provides for accurate representation of decision impacts – for example, the impact of a decision to double the resources applied to a specific strategic investment.

In this step, each strategic investment option is evaluated in terms of the performance metrics. This involves measuring each option's anticipated contribution toward meeting the utility's goals, based on the agreed-upon performance scales, and scoring the options on the basis of their contribution toward those goals. These scores are used to prioritize strategic investment options for the utility's strategic investment portfolio.

Over time, as the strategic planning process matures, the scoring methodology can be validated and refined.

Utility management requires allocation of limited resources to address a spectrum of challenges and opportunities. These choices require managers to make judgments about the relative value of one investment compared with others. Using informal, intuitive evaluation methods, force of personality, or political gamesmanship may dictate investment strategy – and lead to sub-optimal results. This step of the strategic planning process **formalizes the evaluation process** to ensure defensible outcomes that contribute measurably to utility goals.

With carefully developed performance measures and scales, the investment evaluation process is relatively straightforward. Investment option performance is projected, compared against performance scales, and scored. However, actual scoring is typically laden with judgments about both proper interpretations of performance metrics and projected investment performance. Therefore, in practice, the evaluation process provides an opportunity for utility management to galvanize their collective insight on the **relative merits of alternative investment options**. Over time, as the strategic planning process matures, the scoring methodology can be validated and refined.

Key activities for Step 3

The following diagram delineates the activities and decision points recommended for completion of Step 3. This diagram is organized to display the relative sequence of events, from left to right. From top to bottom, the activities are organized by organizational level: policy board activities and interaction are shown at the top, with senior management,

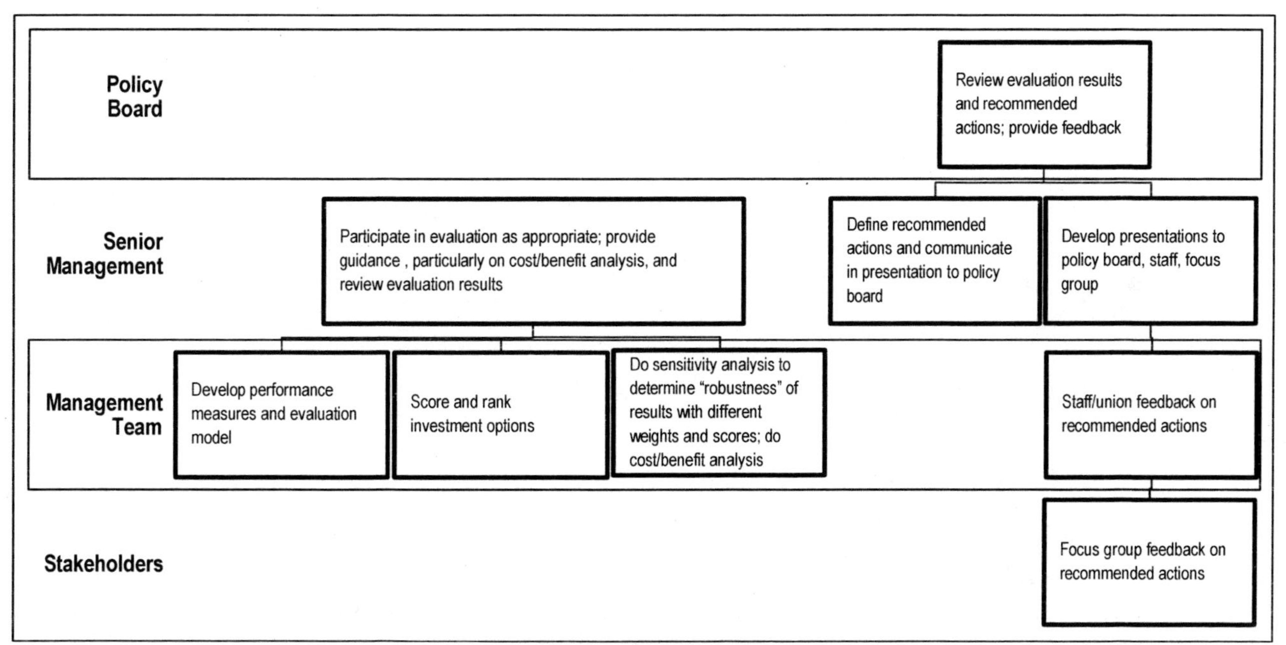

Key activities and decision points for Step 3 – Evaluate Investment Options

utility management team, and stakeholder activities shown beneath. It is generally expected that the activities in Step 3 will take 2 to 3 months to complete, varying with the utility's chosen level of analytical complexity. A description of Step 3 completion for the Honolulu BWS is included on the following pages. Further details on recommended steps for strategic planning will be developed in a follow-on Research Foundation study.

Applicability to water utilities

Utilities entering into a strategic planning process may uncover an alarming lack of available data to measure current performance. Even utilities that have undergone extensive benchmarking efforts may not be collecting data that can show directly how the utility is performing in meeting its fundamental goals. Rather than seeking a multitude of benchmarks that may or may not relate to the utility's goals, the strategic planning process helps focus performance measurement to guide utilities toward goal achievement.

Because utilities may not have data to mark their current positions, the strategic planning process is best considered as an evolutionary process. In particular, strategic planning – and the collection of performance measures to guide it – is appropriately viewed as an **"adaptive implementation" process**, wherein new performance measures may be developed and refined as more and better information becomes available.

Case study description

Findings

- The managerial staff's substantial intuition on performance measures made it possible to construct, revise, and confirm performance scales in a single workshop.
- The absence of available data from which to establish existing conditions made it clear that data collection efforts deserve priority to enable measurement of utility performance. The utility's ongoing business process reengineering needs to ensure collection of, and efficient reporting on, data related to established performance measures.
- The significant portion of the evaluation was accomplished in a 2-day workshop and a series of one-on-one interviews of the operating unit managers. The team reviewed preliminary investment scoring, conducted scenario analyses, constructed portfolios, and examined changes in investment rankings given changes in selected input variables. These outcomes served as the principle basis for drafting the strategic plan.
- BWS management staff easily used draft performance scales and intuition to score strategic investment options. This scoring process was viewed as accessible and required a limited investment of time – generally completed by the consultant within a 2-hour interview of each operating unit manager.

Honolulu Board of Water Supply
Goal: Expand Water Resources
Performance Scale

Score	Description
5	Provides 3+ mgd of potable water supplies OR 1+ mgd of non-potable supplies with a high (50%+) probability of 50+ year sustainability AND no degradation of watershed characteristics.
3	Provides 1-3 mgd of potable water supplies OR 0.5 -1.0 mgd of non-potable supplies with a high (50%+) probability of 25-50 year sustainability AND limited degradation of watershed characteristics.
1	Provides up to 1 mgd of potable water supplies with a high (50%+) probability of 25-50 year sustainability OR substantial degradation of watershed characteristics.

Goal: Become More Competitive
Performance Scale

Score	Description
5	Results in a 10% or more reduction in the gap between BWS performance and established target benchmarks of 2 or more performance measures.
3	Results in a 10% or more reduction in the gap between BWS performance and established target benchmarks of 1 performance measures.
1	Results in less than 10% reduction in the gap between BWS performance and established target benchmarks of 1 or more performance measures.
Benchmarks	O&M Cost / Customer; Number of Staff Positions / Unit of Asset Used; Revenues / $ of Fixed Asset Inventory; Debt Service Coverage; % of staff that define job as advancing BWS goals.

BWS developed a set of performance scales for each of its utility goals. These performance scales used combinations of objective measures and qualitative descriptions of utility performance.

- Despite the absence of detailed data on current performance relative to strategic goals, or refined projections of strategic investment performance, scoring of the benefits of strategic investment options was remarkably consistent across all management staff.
- The potential market conditions identified in scenario planning (during Step 2) proved to be a helpful vehicle for examining the robustness of BWS management staff's scoring of strategic investments.

Experience

The development of performance measures and scales for the BWS's strategic planning process was facilitated by the Board's parallel reengineering efforts. Because the Board's operating units were in the process of evaluating existing and constructing new business processes, BWS leadership was already focused on defining standards of performance – which facilitated the development of performance measurements for the evaluation of strategic investment options.

The strategic planning process also served to focus BWS leadership in a number of different respects. First, whereas the utility's reengineering efforts had been largely focused on detailed examination of individual business processes, this step of the strategic planning process required a broader perspective. The Board's leadership was asked:

- How will the Board determine if it has accomplished its fundamental goals? What measures will evidence success?
- What measures will the Board's stakeholders use to gauge performance?

These questions were oriented toward defining a limited number of goal-aligned performance metrics. Measures focused on overall performance rather than the minutia of efficiencies of specific business processes – in other words, measures that will resonate with the Board's customers, community, and key decision makers.

Second, the development of performance measures for the Board's strategic planning process permitted a focus on the Board's ***strategic investments***. In general, there is the danger that while enhancing the efficiency of current practices, a utility may forgo opportunities to make fundamental changes in its approach to service delivery. The industry's recent drive to achieve "competitiveness" has, at times, translated into self-defeating cost cuts.[4] Defining goal-aligned performance measures helped ensure that BWS leadership prioritized those investments that were anticipated to yield the most significant progress towards the utilities' goals.

---

[4] For example, to the extent that staff reductions have compromised utilities' ability to conduct adequate asset management programs, the near-term benefit of reduced labor expenses may be far outweighed by the failure to renew and replace aging infrastructure sufficiently far in advance to limit service outages and costly emergency repairs.

Third, the exercise of defining performance measures and scales specifically aligned with BWS's goals and objectives highlighted the utility's lack of available data on overall utility performance. This suggested the need to install processes to collect and evaluate high level performance data and adapt performance measures as new information is accumulated.

Natural *scales* refer to direct measures of performance, like elapsed time for customer service responses or survey response percentages.

Constructed scales are detailed descriptions of the qualitative aspects of different levels of performance.

Draft performance measures and scales were developed with BWS staff through a workshop whose outcomes were remarkable in several respects. This workshop introduced the general concept of portfolio management as a basis for strategic planning, outlined techniques for strategic investment evaluation, and defined requirements for performance measures and scales. In these workshops, uses of ***natural*** and ***constructed*** performance scales were demonstrated, in part by providing examples of the types of performance measures that might apply to BWS's approved goals. Noteworthy, and likely not atypical of many utility organizations, were

- Utility management's strong intuition as to appropriate performance metrics and scales ***for their utility*** by which to evaluate strategic investments;
- The relative ease with which participating BWS staff members achieved an understanding of, and comfort with, the process of defining goal-aligned performance measures and scales;
- The limited time required to confirm or revise performance measures and scales – a single workshop session.

Despite the absence of readily available data from which to establish existing conditions, and the need for prospective business process reengineering to ensure collection and efficient reporting on established performance measures, the very process of developing performance measures proved useful in the BWS's subsequent evaluation of candidate strategic investment options.

The evaluation of BWS's strategic investment options was done in a two-step process that included a set of "scoring interviews" with individual BWS managers, and a collective scenario planning/portfolio building workshop. The process was, in some respects, stylized in that BWS had limited data on the utility's existing performance relative to goal-aligned metrics, and had conducted limited research on many of its candidate strategic investments. What transpired, therefore, was that BWS managers scored strategic investments based largely on intuition and by reference to the draft performance measures and scales developed in Step 3. The interviews took less than 2 hours in all cases.[5]

[5] Though strategic investment scoring should be relatively straightforward and unambiguous, it is anticipated that utility managers with larger libraries of data and reports to reference related to candidate strategic investments may require more time to conduct scoring. The time commitment will vary depending on the availability and quality of data sources but is not, in any case, anticipated to be onerous.

Remarkably, scorings across BWS managerial staff were quite consistent, rarely varying by more than 1 point over a 5-point scale. These results suggest that either perceptions of future benefits from strategic investment opportunities are well-defined and understood, or that top BWS management has conveyed a strong sense of commonality of purpose, or both.

Either way, BWS managerial staff's independently consistent evaluation of investment options provided a strong foundation for collective decision making through a single portfolio construction workshop.

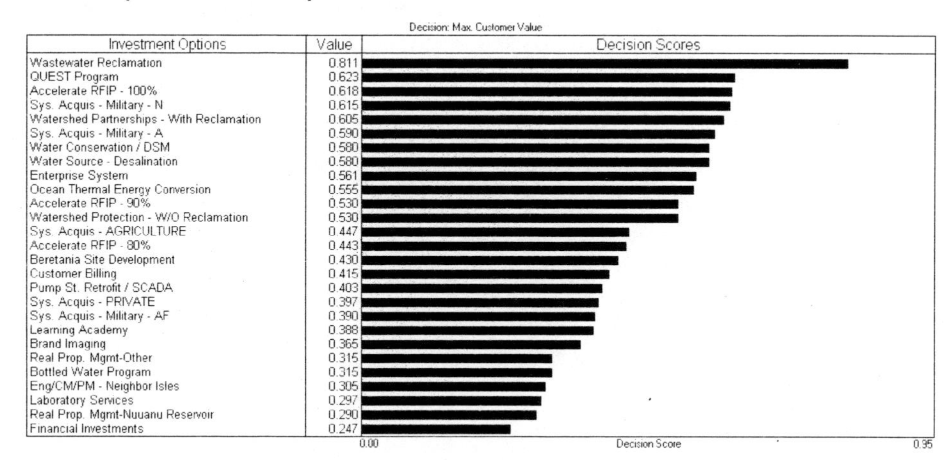

BWS management staff gave similar scores to the investment options. The resulting preliminary ranking provided a strong foundation for decision making.

In reviewing benefit scores across all strategic investment options, it was noted that the scale of these investments ranged dramatically from multimillion-dollar infrastructure investments to relatively limited investments in enhanced community communications and involvement. As with personal financial portfolio management in which the investor selects from alternative classes of investments – real estate, stocks and bonds, and durable goods – a "portfolio" of utility investments will draw from alternative classes of investment options. A simple classification system employed for the BWS case study was to class strategic investments by cost category. Classes of strategic investment options, ranked by benefit score, served as the basis for construction alternative investment portfolios in the next step of the strategic planning process.

Step 4: Develop strategic plan

Purpose of this step

While preparing a strategic plan offers obvious benefits in terms of documenting the outcome of the strategic planning process, the true purpose of this step is to require that fundamental decisions related to the utility's strategic direction be made and communicated to internal and external stakeholders. By developing and publishing a strategic plan, utility management and stakeholders are required to select specific strategic investments and identify how these investments are aligned with utility policy goals.

Description

Simply put, the development of a strategic plan may involve determining the number and type of high-ranking strategic investments that may be made within the utility's financial constraints. The strategic plan is thus a high-level, accessible summary of the utility's annual budget.

The principal value of strategic planning, however, is generally better realized when utility management critically consider how strategic investment options may advance the utility's fundamental goals and objectives. It is utility management, more than any other party, that understand and appreciate the

- Complexities involved in effecting strategic investments,
- Potential synergies across investment options,
- Constraints on timing and execution of available investments, and
- Delicate balance between daily operations and new initiatives.

This critical thinking may be facilitated by developing alternative portfolios of strategic investment options that the utility may decide to pursue. In the same way as personal investments may reflect greater or lesser appetite for risk, preferences for real properties versus securities, and so on, alternative utility investment portfolios may reflect a greater focus on entrepreneurial opportunities, asset management, or human resource development.

In the same way as personal investments may reflect individual preferences, ***alternative utility investment "portfolios"*** may reflect a greater focus on entrepreneurial opportunities, asset management, or human resource development.

Procedurally, building portfolios generally involves examining the rankings of different classes of strategic investment options accomplished in Step 3. Different portfolios of candidate investments may be built, typically reflecting competing priorities, windows of opportunity for selected investments, and synergies across investment options. **Scenario planning** may be used to help build portfolios that are responsive to potential changes in business conditions.

Once alternative portfolios of investment options are built, utility management (in conjunction with internal and external stakeholders) evaluates these complete portfolios in much the same way as the individual investment options were evaluated. That is, portfolios are ranked based on their projected performance relative to the utility's fundamental goals and objectives. These rankings may be tested by reference to scenario planning efforts to ensure that the selected portfolio will be robust given potential changes in business climate.

From this point, developing the utility's strategic plan is a matter of documentation and communication. The formality of documentation and extent of communication will reflect the cultural and political environment within which the utility operates. Notably, several of the most successful utilities in the water industry (such as Seattle Public Utilities and Louisville Water Company) use their strategic plans as prominent communications tools both within and outside their utilities. (See Chapter 3 for more information on communications planning.)

In reviewing strategic plans from utilities both within and outside the water utility industry, several common attributes were found. In particular, most of the printed summaries of strategic plans were written in accessible language, supplemented by compelling graphics, and included:

- Introduction or message from executive
- Mission, vision, and goals statements
- Review of marketplace challenges and opportunities
- Review of selected strategic investment options
- Basic information on the utility organization and financial condition

Key activities for Step 4

The following diagram delineates the activities and decision points recommended for completion of Step 4. This diagram is organized to display the relative sequence of events, from left to right.

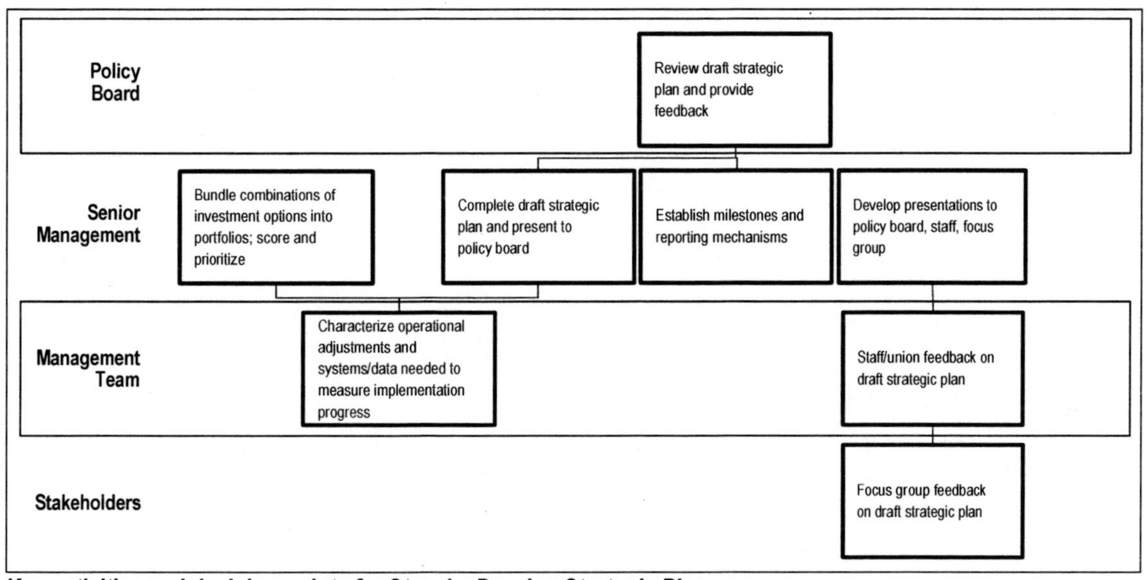

Key activities and decision points for Step 4 – Develop Strategic Plan

From top to bottom, the activities are organized by organizational level: policy board activities and interaction are shown at the top, with senior management, utility management team, and stakeholder activities shown beneath. It is generally expected that the activities in Step 4 will take 2 to 3 months to complete, varying with the utility's chosen level of analytical complexity. A description of Step 4 completion for the Honolulu BWS is included on the following pages. Further details on recommended steps for strategic planning will be developed in a follow-on Research Foundation study.

Applicability to water utilities

Throughout much of the history of water utility operations in North America, there has been little need for utility managers and stakeholders to contemplate market strategy, much less develop strategic plans. However, as the utility market has changed in the last decade, **the very survival of individual utilities compels the need for strategic thinking** and more effective communication of market strategies. Utilities that are unable to adapt to changing market conditions and proactively address prevailing market challenges will face consolidation.

The actual development of a strategic plan provides a format for a utility to examine its market position, as well as its constraints and opportunities, and to galvanize staff and stakeholders around a defined strategic direction. Utility management may articulate the implications of constraints on its freedom to act and define how it will prioritize uses of resources to accomplish established goals and objectives. The process itself forces utilities to access the implications of changing market conditions and determine a proactive strategic direction. Whether conducted rigorously, with substantive evaluation of options based on projected performance, or as a structured managerial exercise, **strategic planning is a mechanism for all utilities to better position themselves for success** in a competitive marketplace.

Case study description

Findings

- BWS management staff, though relatively unfamiliar with concepts related to portfolio management and scenario planning, successfully engaged in the classification of investment options and development of scenarios. Given limited facilitation, the concepts of portfolio management and scenario planning proved accessible to utility management.
- Unsurprisingly, given the relative consistency of individual management staff scoring of investment options, there was general alignment of strategic plan elements with new utility initiatives defined by the BWS's new leadership and affirmed via its reengineering program.
- Integration of strategic plan outcomes into the annual planning cycle requires timely communication to internal and external stakeholders, and sustained commitment of utility leadership and management. The value of the strategic planning process for BWS was limited by distracting parallel reengineering efforts and urgencies presented by emerging market opportunities.
- The degree of ownership of the strategic plan that BWS managerial staff exhibited was directly associated with their level of participation in and commitment to the strategic planning process.

Experience

BWS managerial staff participated in a set of strategic planning workshops, conducted after each strategic investment option was scored individually through one-on-one interviews. These half-day workshops, conducted in September 2001, included a presentation and discussion of scoring results, classification of strategic investments by level of resource commitment, and a scenario planning exercise.

Procedurally, the "raw" benefit scores assigned by individual BWS managers and averaged across the management team were reviewed, and significant differences in scores were discussed. This led to a fuller understanding of BWS management perspectives on the potential performance of candidate strategic investments. A baseline portfolio was constructed by identifying the top two to four ranked strategic investment options in each class of investment, based on average scores of BWS managerial staff. From this baseline, selected revisions were made, via consensus of management staff, to ensure that potential synergies across different classes of strategic investments were realized. As such, a strategic investment portfolio was established.

Scenario planning was conducted largely to test the robustness of the investment portfolio. Three general scenarios were discussed, including the occurrence of favorable and unfavorable business conditions, as well as preservation of the status quo. Notably, this discussion was punctuated by consideration of the ramifications of the terrorist acts of September 11, 2001, which had significantly reduced tourism in Hawaii. In general, scenarios were described by reference to potential changes in economic, political, social, and environmental conditions that could affect BWS's business opportunities.

Honolulu Board of Water Supply
Summary of Scenarios for Scenario Planning

Scenario	Favorable	Status Quo	Unfavorable
Economic	Asia economy improves and mainland economy begins to recover. High-tech or other manufacturing or service center enterprise look to Honolulu as center for operations.	Slow growth. Service sector dependent. Rates and charges under pressure for no increase.	Asian economic decline accelerates and spreads. Mainland economy continues to decline or remains lethargic.
Political	Support of BWS quality service is recognized. Autonomy is favored. One-utility concept is embraced.	Stable political leadership. High level of public confidence. Autonomy of BWS is emerging.	Political situation becomes unstable. Change in leadership at city. Privatizers win attention of political leadership.
Social	BWS is seen as provider of quality services and major contributor to quality of life in Honolulu.	Water supply alternatives remain controversial. Hawaiian rights are actively expressed in water management policies.	Hawaiian rights movement intensifies.
Environmental	Watershed rehabilitation is successful. Weather conditions remain stable; no hurricanes or droughts.	Watershed management and protection of natural streams and rivers is a priority for environmental activists and Hawaiian culture. Basin transfers are highly controversial.	No-growth advocacy intensifies and limits expansion of system. Islands enter into long-term drought.

Other considerations involved in evaluating the top-ranking strategic investments related primarily to timing, logistics, and institutional concerns. For example, while acquiring selected military facilities' utility systems was highly ranked, it was recognized that BWS could not dictate the timing of investment in this opportunity. Similarly, while multiple reengineering and enhanced information management systems scored highly, it was recognized that concurrent information management system conversions likely was impractical.

Notably, a general consensus regarding the prioritization of candidate strategic investments was accomplished with relative ease, and was readily demonstrated to be in alignment with the utility's established goals and objectives. The drafting of a strategic plan consistent with the affirmed portfolio of strategic investments was accomplished without significant difficulty. In addition, while the merit of most priority strategic investments was affirmed, the prioritization process also demonstrated that selected "favored" strategic investments did not merit focused staff attention.

"The Honolulu BWS's strategic plan demonstrates an advanced sense of ***'community as customer'*** – the quality-of-life impacts of its decisions, and the economic effects on the community at large."

Diana Gale
Seattle Public Utilities

While the draft strategic plan developed in December 2001 reflected a general consensus on the strategic direction of the Board, formal review, revision, and adoption by the BWS management and Board was suspended throughout the majority of 2002. As such, some momentum for including strategic planning in the Board's annual planning cycle was forfeited, and some distractions limited progress on strategic investments. Accordingly, case study experience also affirmed the importance of maintaining adequate leadership and commitment to the strategic planning process. To be effective in guiding its strategic direction, a utility must **incorporate the tenets of its strategic plan into ongoing management of utility activities and initiatives**. This is facilitated by detailed implementation planning, done in the final step of the strategic planning process.

Successful strategic planning requires continued leadership and commitment, balancing the flexibility to respond to change with the stability to weather shifts in political winds.

Case study experience also confirms that **strategic planning, and its integration into utility activities, is a *continuous* and *long-term* process**. It may take time to realize results. The only certainty is that successful strategic planning requires continued leadership and commitment, balancing the flexibility to respond to change with the stability to weather shifts in political winds.

Step 5: Plan for implementation

Purpose of this step

Strategic planning is, by definition, a relatively high-level activity that addresses the strategic direction of the utility. However, for this planning to have value, the specific tactical requirements to effect strategy must be addressed. In the absence of planning and effective management of strategic investments, those resource allocations deemed to be of utmost importance for accomplishing the utility's fundamental goals and objectives may be squandered. Planning for implementation defines the scheduling, budgeting, communications, and work processes required to ensure that projected performance of strategic investments are realized. Implementation planning helps ensure new initiatives are coordinated with daily service delivery responsibilities, defines appropriate expectations regarding progress towards objectives, and translates often lofty goal statements into calls for deliberate action, with defined assignments.

Strategic planning translates often lofty goal statements into calls for deliberate action, with defined assignments.

Description

Planning for implementation involves an exacting assessment of what resources will be required to make strategic investments. Similar to program or project management, it involves establishing specific work plans, assignments, schedules, and budgets for individual strategic investments. Interdependence of related tasks, potential duplication of claims on available resources, and institutional constraints must be addressed.

Implementation planning generally employs a variety of well-established tools and techniques that enable effective management of available resources within prevailing constraints. Cosmetically, it may be accomplished using highly sophisticated software tools with elaborate graphics. On the other hand, using mundane calendars, budgets, and status reporting approaches can work just as well depending on the complexity of implementation requirements. Fundamentally, implementation planning must provide mechanisms by which utility management may define and monitor how strategic investments will be effected for the organization.

Strategic investments, however, may present atypical challenges to implementation planning – particularly those investments that represent entrepreneurial efforts beyond the utility's historical purview. While water utilities are practiced in infrastructure project delivery and water utility customer services, implementation planning that requires assessments of market conditions, potential competitor analyses, development of marketing strategies, and so on present new challenges.

Similarly, the metamorphosis of the water utility industry's market conditions has not necessarily been accompanied by parallel changes in the institutional and legal frameworks within which many utilities operate. Unlike other actors in a competitive marketplace, **water utilities must address planning for redress of institutional and legal barriers** to their participation in various forms of enterprise.

Key activities for Step 5

This diagram includes the activities recommended for completion of Step 5. This diagram is organized to display the relative sequence of events, from left to right. From top to bottom, the activities are organized by organizational level: policy board activities and interaction are shown at the top, with senior management, utility management team, and stakeholder activities shown beneath. It is generally expected that the activities in Step 5 will take approximately 1 month to complete. A description of Step 5 completion for the Honolulu BWS is included on the following pages. Further details on recommended steps for strategic planning will be developed in a follow-on Research Foundation study.

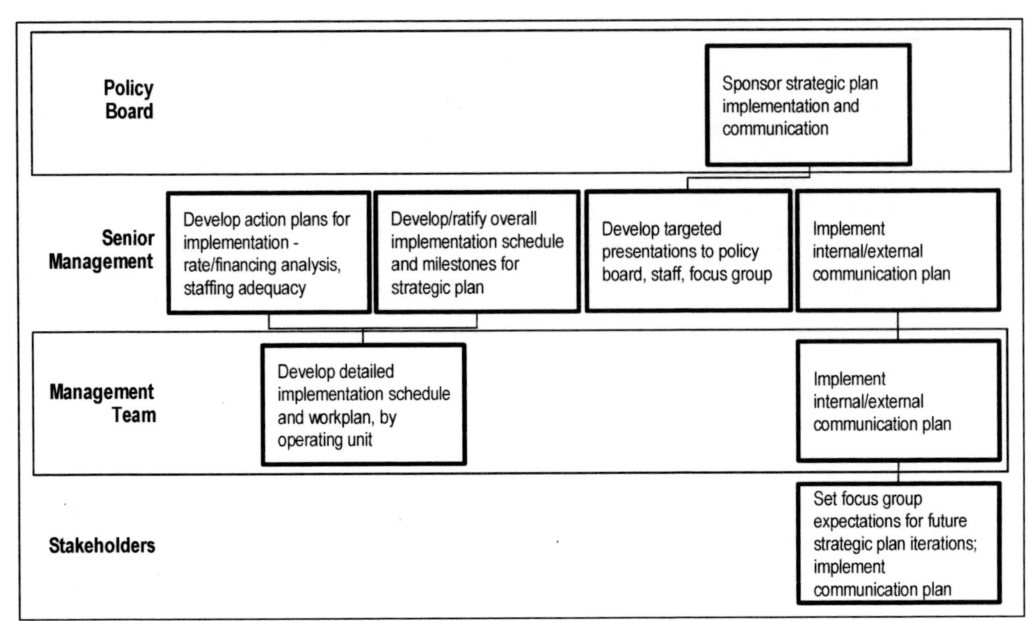

Key activities and decision points for Step 5 – Plan for Implementation

Applicability to water utilities

Because strategic planning has not historically been required of most water utilities, and because simple preservation of the status quo may go unchallenged, strategic planning efforts typically have proven more ornamental than functional. Planning for implementation, and imposing a strong mandate for compliance with these plans, is required for utilities to proactively define their position in the changing utility marketplace rather than assume a reactive posture.

Implementation planning is the vehicle by which ***positive, deliberate movement forward on strategic investments*** can be integrated into the utility's operations – allowing the utility to advance, while not compromising "mission critical" performance.

Implementation planning for strategic investments is absolutely critical for sustaining the strategic planning process and making it *real* for the utility's internal and external stakeholders. While general utility goals and objectives are meant to convey a vision of future utility performance and strategic direction, daily imperatives of protecting public health, reliable customer service, and environmental stewardship may easily consume attention and energies. The requisite juggling of priorities is difficult – even more so while moving forward. Implementation planning is the vehicle by which positive, deliberate movement forward on strategic investments can be integrated into the utility's operations – allowing the utility to advance, while not compromising "mission critical" performance.

Case study description

Findings

- BWS management staff readily constructed implementation plans with quarterly milestones and performance metrics for each of the strategic investments they were assigned to address. These plans reflected an appreciation of existing institutional constraints and requirements to balance strategic investments with daily operational needs.
- A common business plan template that requires delineation of milestones, deliverables, performance metrics and resource requirements (see Appendix B) enabled standardized reporting on strategic investments and may serve as a readily updated management tool.
- The BWS's deferral of its publication of the strategic plan and follow-up to initial implementation planning reinforces the importance of these aspects of the planning process. For the strategic plan to be successfully integrated into the utility's annual planning processes, ensuring accountability for achievement of defined milestones is essential.

Experience

In parallel with completion of the draft BWS strategic plan as described in Step 4, BWS management staff participated in a series of one-on-one interviews to complete a standardized business plan template. This template, provided in Appendix B, required BWS management staff to

- Articulate operating units' strategic objectives for the forthcoming year, and for the 5-year forecast period;
- Indicate how these objectives will contribute to achievement of the utility's overall goals and objectives;
- Define specific performance metrics for each objective and the relation of these focused performance metrics to those established for the utility's overall goals;
- Catalogue general requirements for successful implementation by indicating whether the investments will require business process reengineering, information technology, organizational development, public communications, staff training, and cross-operating unit coordination; and
- Provide general estimates of operations and maintenance and capital expenditures.

"The business planning template is a ***powerful tool***; it recognizes the importance of performance measures, and it drives goal attainment down into the utility."

Mark Premo
Anchorage Water and Wastewater

Once oriented to the requirements of the business plan template, BWS management staff were able to substantively complete the exercise in the course of an intensive 2- to 3-hour interview session. Based on concluding comments of participating staff, as well as top BWS management, this process proved particularly valuable because it forced the parsing of major initiatives into tasks that could reasonably be accomplished within a 3-month time frame. As such, **the daunting**

challenges of strategic investments were dissembled into specific steps that could be taken readily within the constraints of daily operating imperatives.

The BWS deferred revision of its draft strategic plan and publication thereof, as well as use of operating units' implementation plans to direct and monitor strategic investments. This deferral was occasioned by competing urgencies, as well as continued organizational reengineering. However, integration of strategic planning and focus on certain of the utility's strategic investments has been truncated. The utility decided to revitalize the strategic planning process for fiscal year 2003 planning, which required specific additional training and implementation efforts. This outcome highlights the importance of integration of strategic planning processes into annual planning efforts and ongoing management focus to ensure the sustainability and relevance of strategic planning outcomes.

Chapter 6: Conclusions

The dynamic pace of the water utility industry is challenging utility managers more than ever to perform efficiently, optimize resource investments, and engage their customers and communities. This challenge is by no means simply a call to achieve available savings in operating costs and project delivery. Instead, utilities increasingly are being required to integrate, and effectively balance, resource investments across competing sets of goals. For utilities to succeed in this environment, they must define a strategic direction that realizes available opportunities within the constraints of their freedom to act.

Recommended approaches to strategic planning recognize that water utilities' first and foremost obligations are the provision of reliable and safe water to their customers. Therefore, strategic investments are defined as those resource allocations – above and beyond those required to meet operational imperatives – that enhance existing customer services, improve risk management, or expand service offerings. In effect, strategic investments define new directions that take advantage of the opportunities presented by marketplace dynamics. Planning for these new directions should begin with affirming or updating the utility's goals, as appropriate, and should be integrated into the utility's annual planning cycle so that implementation of strategic investments is balanced with requisites of operational imperatives. In this manner, utility managers will be able to place siren calls for asset management, enterprise information management, total watershed management, and other strategic initiatives – within a cohesive strategy.

At the Honolulu Board of Water Supply, despite complications associated with a parallel reengineering effort and limitations on available utility performance data, strategic planning demonstrated the value of structured decision processes for strategic planning, and the applicability of private-sector strategic planning principles to water utilities.

This case study experience, in combination with research conducted through participation of the project Applications Board, also highlighted the importance of sensitivity to the political environment in which the utility operates, as well as its policy board's and stakeholders' appetite for risk and change. Comprehensive strategic planning requires a realistic assessment of the ability to secure political acceptance for, and actually implement, strategic investment options. In many cases, particularly for utilities with less mature planning processes and information management systems, adequate information to support rigorous strategic planning may not be available. As a consequence, utilities may initiate strategic planning by "planning to plan" through implementation of business process analyses and performance monitoring to better characterize available opportunities.

For strategic planning processes to yield value, they must be sustainable and effected over multiple years as utilities evolve their organizations to succeed in their changing environment. While convenient tools may be used to translate strategic direction into tactical action, the sustained commitment of utility leadership and policy boards is of paramount importance. Utilities face unprecedented challenges in the new millennium. Strategic planning provides a framework to achieve success by ensuring that resource allocation decisions are directed toward fundamental goals – that is, by ensuring that utility managers and decision-makers "keep their eyes on the prize."

Ongoing research

This tailored collaboration project represents another important step[6] in the development of effective tools and processes to meet the challenges of the changing utility marketplace. Notably, it demonstrates the potential application of techniques used in the private sector to address water utility management challenges. One of the Research Foundation's projects slated for FY 2002-03, *Strategic Planning and Organizational Development for Water Utilities,* will continue this line of research and process development. These projects will provide a durable and overarching framework for utilities' discrete planning functions, with the optimal result of realizing the industry's movement towards truly integrated planning.

[6] This project builds on previous Research Foundation projects related to public involvement strategies, evaluation of public-private partnership options, capital planning strategies, and utility futures. For a complete listing of Research Foundation projects related to utility management challenges, visit the AwwaRF website at <www.awwarf.org>.

Appendix A:
Draft Strategic Plan for Honolulu BWS

DRAFT

Tuesday, December 18, 2001

HONOLULU BOARD OF WATER SUPPLY

STRATEGIC PLAN

FY 2002-03

Pure Water

- ***Our Greatest Need***
- ***Use It Wisely***

Our Community

The Board of Water Supply serves a diverse and active stakeholder community – from our forward-thinking Board of Directors to our watershed management partners, from our fellow government agencies to individual customers and community organizations.

We work collaboratively with each of these stakeholder groups to ensure that we understand and address their concerns and that we tailor our services to meet their needs.

Stakeholders

- Our customers
- Our employees
- Island visitors
- Military installations
- Oahu residents
- Board of Directors
- State of Hawaii
- City and County of Honolulu
- Hawaii Department of Health
- Hawaii Commission of Water Resource Management
- Cultural organizations
- Neighborhood organizations
- Environmental organizations
- Agricultural interests
- Development interests
- Industrial interests
- The environment of our island

Message from the General Manager

Over the last several years, we at the Honolulu Board of Water Supply have recognized that the water services marketplace is becoming increasingly complex, dynamic, and competitive. We are being asked to do more ... and we are meeting and exceeding the challenge.

We have restructured ourselves to become more competitive, changing our tools, techniques, and work processes to enhance efficiency and improve service quality. Now we seek to not only meet our customers' needs with efficiently delivered services, but to anticipate their future needs.

We have assumed a leadership role in the environmental stewardship of Oahu's precious natural resources. We are forming community partnerships to protect our watersheds and are working with the City of Honolulu and State of Hawaii to effectively manage our water resources. We are developing water supplies that will afford a foundation for sustainable economic development.

We have expanded our breadth of services to leverage the exceptional capabilities of our staff and take advantage or emerging market opportunities.

Perhaps most importantly, we have effected change while acknowledging that our unique history, culture, and people are our greatest source of strength. We celebrate BWS's *ohana* as we prepare to meet the challenges of the 21st century.

This Strategic Plan provides a foundation for our continuing evolution, guiding our strategic investments and maintaining focus on our strategic goals. It presents our Mission, Vision and Strategic Goals; characterizes our operating environment and marketplace; and identifies the strategic investments and initiatives that will enable our continuing success. This Strategic Plan will be a 'living' document, charting BWS's path forward while honoring our people and past.

- Cliff Jamile, General Manager

Our Mission

We will improve the quality of life in our community by providing world-class water services.

Our Vision

We will share our water management expertise and provide services statewide and throughout the Pacific to:

- Benefit our global communities.
- Benefit our community by maintaining reasonable rates.
- Provide professional development opportunities for our staff.
- Support the growth of Hawaii's economy.

Our Values

We will honor our people, our culture, our environment, and our community by:

- Striving for excellence in our performance
- Innovation in our tools and processes
- Integrity in our conduct

Our Strategic Planning Process

The Honolulu Board of Water Supply (BWS) has initiated an annual strategic planning process to help ensure that our resource investments are directed to best serve our ratepayers, environment, and community. This process begins with affirmation of our Strategic Goals, reviews our operating environment and constraints, and identifies opportunities for BWS to enhance service to our community. Strategic investments are then prioritized based on their anticipated contribution to the achievement of BWS goals.

This strategic planning process provides BWS a foundation for development of our annual operating and capital budgets. It defines a 'portfolio' of strategic investments that are aligned with our strategic goals and will enable BWS to both proactively shape our services to address our customer's needs, and continue to steward our island's precious natural resources.

Our Strategic Goals

- **Expand Water Resources**
- **Become More Competitive**
- **World-Class Customer Service**
- **Improve Infrastructure Reliability**
- **Develop Staff and Renew BWS *Ohana***
- **Identify Opportunities for Growth**

Our Strategic Goals define what we seek to accomplish to fulfill the BWS Vision. They reflect the unique challenges that BWS faces as Oahu's water services provider – simultaneously addressing infrastructure development needs, opportunities to enhance customer services, and responsibilities for stewardship of our environmental resources.

Assumptions

Our approach to accomplishing these Strategic Goals is framed by the institutional, legal, and market conditions that define our opportunities to change, grow, and improve. We have assessed these conditions and developed our strategic plan under the following assumptions related to our strategic goals:

Water Resources

- Groundwater supplies will continue to be our primary water source, requiring our protection of quality and sustainability.
- As groundwater is a limited resource, there is a need to develop future water supplies. Research and piloting of desalination technologies will enable our development of a permanent water supply resource.
- We will exercise leadership in the stewardship of Oahu's water resources – protecting resources from mountains to oceans.
- BWS will avoid inter-district transfers of water.

Competitiveness

- We will continue to operate in a competitive utility market, leveraging opportunities and meeting challenges.
- Collaborative relations with labor representatives will be critical to enhancing our efficiency and effectiveness.
- Expansion of our scope of services will require entrepreneurial investments and effective risk management.

Customer Service

- Our customers' service expectations will continue to heighten as technology and other service sectors advance.
- We will deliver enhanced customer services with rate increases that do not exceed increases in the Consumer Price Index.
- Our customers' confidence will be impacted by the relative stability of international politics and economies, and by BWS's record of regulatory compliance.

Infrastructure Reliability

- Proactive renewal and rehabilitation of our fixed asset base will determine the long-term reliability of our services.
- New technologies and maintenance management practices will enable extension of asset service lives.

Develop Staff and Renew BWS *Ohana*

- Continuous improvement of our skills and knowledge base are required to maintain competitiveness and enable expansion of services.
- We will empower employees through team building, training, and alignment of responsibilities with decision-making authority.
- We will continue to invest in employee training and to challenge our employees with new roles and responsibilities.

Business Growth

- We will expand our products and services with respect for boundaries of public enterprise.
- We will develop opportunities to broaden the scope of BWS services to leverage available economies and enhance water resource stewardship.
- We must be prepared to respond quickly and efficiently to scheduling of business growth outside of our control.

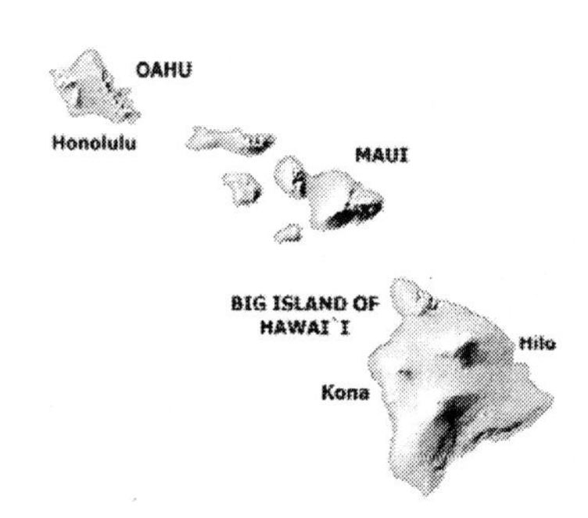

Our Market Position

BWS is uniquely positioned in the water services marketplace – a market that extends across the entirety of the Pacific Rim – to deliver needed services. While substantial challenges await us, we have the ability to realize our Vision.

Our Strengths

- BWS's *ohana*
- Board and community support
- Experienced and entrepreneurial management
- Established reputation for quality service delivery
- Positive labor relations
- Available employee resources
- Technical expertise
- Financial soundness and self-sustainability
- Competitiveness initiatives
- Emerging community partnerships

Our Challenges

- Overcoming organizational resistance to change
- Institutionalizing community involvement
- Updating information management systems
- Establishing 'brand image' among general population
- Improving job skill training and career track options
- Gaining experience in entrepreneurial investments and risk management
- Updating administrative and customer service offices
- Addressing declining groundwater quality in an uncertain regulatory context
- Achieving community consensus on water resource management issues
- Overcoming resistance to use of reclaimed water and water conservation
- Operating in a climate of global economic and political instability
- Addressing security issues

Our Opportunities

- Emerging availability of utility systems for acquisition
- Pacific Rim market potential
- Availability of information management systems to address enterprise-wide needs
- Potential partnerships with State and community groups to share environmental stewardship responsibilities

Our Strategic Investment Portfolio

Our assessment of market position has framed our evaluation of opportunities to enhance services and steward environmental resources. We have selected a set of strategic investments to better serve our customers and community⁘.

Wastewater Reclamation

In January 2000(?), the Board of Water Supply invested in the Honouliuli Wastewater Reclamation facility, committing to market up to 10(?) mgd• of reclaimed wastewater effluent. The Board's entry into the reclaimed water market represents our single most significant strategic investment, being a major expansion of BWS services and an important step in defining a sustainable mix of water supply options for Oahu.

Our continuing investment in wastewater reclamation will involve:

- Contracting and extension of distribution lines to golf courses located in Ewa.
- Public communications on the economic and environmental benefits of reclaimed water use.
- Proactive marketing and contract negotiation for economically viable reclaimed water sales

Key Performance Measure:

- Contracted Reclaimed Water Sales (Revenue, MGD)

Research and Facility Improvement Program (RFIP)

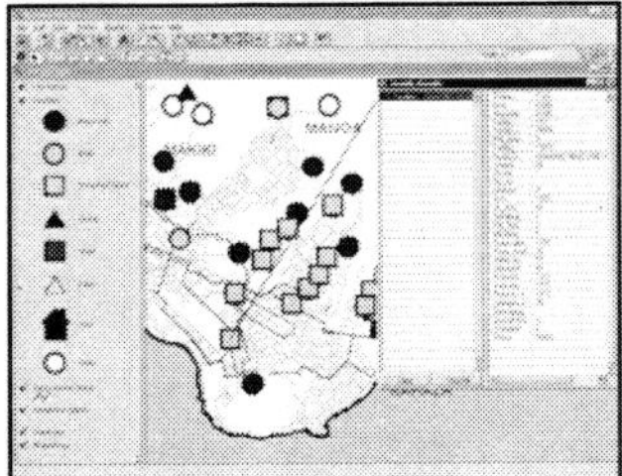

Over the last _ years, we have initiated a proactive asset management program that serves as a benchmark for the water utility industry. We have inventoried our fixed assets, performed condition assessments, and rigorously prioritized renewal and replacement projects. Based on this work, and recognizing the (nationwide) imperative for re-investment in our aging infrastructure, renewal and replacement capital project spending has increased from $__ in 199_ to $__ budgeted in FY 2000-01.

We have elected to continue to re-invest in our existing system assets by:

- Improving our tracking and evaluation of asset conditions, incidences of failure, and contractor performance
- Unifying our evaluation and prioritization of candidate renewal (RFIP) and capital (CIP) projects
- Completing renewal and replacement of an average of __ miles per annum of water transmission and distribution pipelines (subject to unified prioritization)

⁘ Strategic investments are listed in order of greatest potential benefit in terms of accomplishment of strategic goals as determined through the Board's Strategic Planning Process.

• Million Gallons Per Day

Key Performance Measure:

- Percent change in asset failure rates

Watershed Partnerships

Affirming our commitment to collaborative stewardship of our water resources, we will establish watershed partnerships, featuring where economically justified wastewater reclamation projects, to enhance protection of our watersheds.

We will work with the State, City, neighborhood, Native Hawaiian, and environmental groups to coordinate our investments and galvanize community commitment. We will provide leadership by:

- Completing an Watershed Protection Master Plan for Oahu
- Funding watershed assessment studies for major watersheds on Oahu
- Investing in local watershed protection projects (e.g., reforestation, fencing, gauging, water use & source inventory) and reclamation facilities
- Establishing liaison and communications protocols for stakeholder involvement related to all island watersheds.

Key Performance Measure:

- Variance from targeted improvements in water quality indices for selected watersheds

System Acquisitions

As regulatory and infrastructure renewal requirements challenge the economic viability of small, private, agricultural and military utility systems, BWS may be in a position to acquire these systems under mutually beneficial terms.

System acquisitions can advance our role as the premier water services provider in Hawaii, extending services to new customers and enhancing our resource management capabilities.

We will investigate the potential for acquiring utility systems by:

- Developing a protocol for evaluation of candidate system acquisitions
- Monitoring and responding to opportunities for acquisition of military utility systems
- Engaging in discussions with State, agricultural, and other private system interests to prioritize system acquisition opportunities

Key Performance Measure:

- Projected 20-year net present value of system acquisitions

QUEST Program

Over the past several years, we have made an unprecedented investment to restructure our organization and work processes. We have implemented changes and begun business initiatives to enhance competitiveness, improve customer service, and develop staff resources.

We will continue to invest in the QUEST Program by:

- Securing agreements with labor unions for implementation of multi-skill work force and business unit restructuring
- Conducting customer satisfaction and perception surveys and focus groups
- Collecting performance metric data, defining benchmarks, and establishing performance targets

Key Performance Measure:

- __% increase in overall customer satisfaction ratings

Information Systems

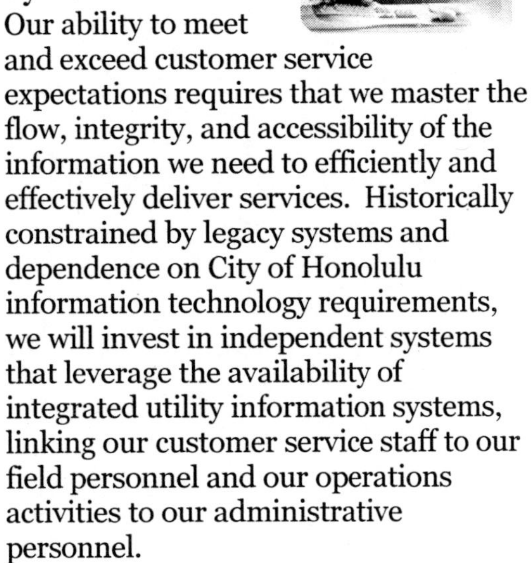

Our ability to meet and exceed customer service expectations requires that we master the flow, integrity, and accessibility of the information we need to efficiently and effectively deliver services. Historically constrained by legacy systems and dependence on City of Honolulu information technology requirements, we will invest in independent systems that leverage the availability of integrated utility information systems, linking our customer service staff to our field personnel and our operations activities to our administrative personnel.

We will enable efficient information management by:

- Developing a flexible, comprehensive IT Master Plan
- Specifying and procuring Customer Information and Financial Accounting Systems that are (or may be readily) integrated
- Enhancing our computer skills training to ensure acceptance and full functionality of newly acquired systems

Key Performance Measure:

- Percentage of staff employing Enterprise System technologies in performance of daily tasks

Water Conservation

Historically, the Board has conducted conservation education and implemented relatively 'passive' programs. Now, with the availability of new programs and technologies to manage customers' water usage, we can expand our water conservation efforts and, in so doing, improve both our service to the community and the management of our water supplies.

We will expand our water conservation program by:

- Implementing conservation-oriented water rate structures for non-residential customers (?)
- Implementing Industrial and Commercial customer conservation retrofit programs
- Employing new analysis techniques to assess water conservation potential and forecast demand reductions

Key Performance Measure:

- __% decrease in average and peak period water use per account (by customer class)

Brand Imaging

Though we enjoy notable community support based on our reputation for service quality, as we improve and expand our services we need to heighten awareness of what we do, our vision, and our commitment to excellence. By promoting greater public understanding of our mission, we can gain support for important programs that will enhance the quality of life in our community.

We will invigorate our community presence through 'brand imaging' efforts that will:

- Highlight our vision, services, and environmental stewardship through broadcast and print media and increased visibility related to community activities
- Revise existing and develop new informational materials on BWS's service offerings

Key Performance Measure:

- _% increase in awareness of BWS services and community role (as indicated through customer surveys)

Real Property Management

As we drive toward world-class performance and reinforce our community image, we will expand our services through active management of our real property holdings. We will begin by revitalizing our administrative offices and service yards to enable our use of the latest technologies and enhance our work environment.

We will communicate, through our physical presence in the community, our commitment to excellence by:

- Redeveloping the Beretania complex
- Evaluating options for redevelopment of other BWS properties
- Evaluating public-private partnership options related to property management services

Key Performance Measure:

- __% satisfaction of employees and customers with BWS offices

Financial Management

Our financial strength and self-sufficiency are critically important to realizing our vision. As we embrace entrepreneurial opportunities, we will:

- Proactively manage our cash reserves and securities holdings
- Redefine our debt and risk management policies and procedures

Key Performance Measure:

- Earned returns on liquid assets

Appendix B:
Business Planning Template

HONOLULU BOARD OF WATER SUPPLY		
STRATEGIC BUSINESS PLAN		
OPERATING UNIT PLAN		
Operating Unit:		
Mission		
Vision		
Values		
Strategic Goals		
Operating Unit: Strategic Objectives		
	Fiscal Year 2001 - 2002	Fiscal Year 2002 through 2006
1		
2		
3		
4		
5		
Operating Unit Objectives: Contribution to BWS Strategic Goals		
Expand Water Resources		
Become More Competitive		
World Class Customer Service		
Improve Infrastructure Reliability		
Develop Staff		
Achieve Business Growth		

HONOLULU BOARD OF WATER SUPPLY

STRATEGIC BUSINESS PLAN

OPERATING UNIT PLAN

Operating Unit:	0

Operating Unit: Strategic Objectives

	Fiscal Year 2001 - 2002	PERFORMANCE MEASURE(S)
1		
2		
3		
4		
5		

	Fiscal Year 2002 through 2006	PERFORMANCE MEASURE(S)
1		
2		
3		
4		
5		

Operating Unit Objectives: Contribution to BWS Strategic Performance Measures

Objective	
Expand Water Resources	
Become More Competitive	
World Class Customer Service	
Improve Infrastructure Reliability	
Develop Staff	
Achieve Business Growth	

HONOLULU BOARD OF WATER SUPPLY

STRATEGIC BUSINESS PLAN

OPERATING UNIT PLAN

Operating Unit: 0

Operating Unit Investments

Required Investments

	Description	Contribution to Operating Unit Goals
1		
2		
3		
4		
5		

Optional (Strategic) Investments

	Description	Contribution to Operating Unit Goals
1		
2		
3		
4		
5		

DISINVESTMENTS (REDUCTIONS IN EXPENDITURES)

	Description	Contribution to Operating Unit Goals
1		
2		
3		
4		
5		

HONOLULU BOARD OF WATER SUPPLY

STRATEGIC BUSINESS PLAN

OPERATING UNIT PLAN

Operating Unit: 0

Operating Unit Investments

Required Investments

	Description	Business Process Reengineering	Information Technology	Organizational Development	Public Communications	Staff Training	Cross Operating Unit Coordination
1							
2							
3							
4							
5							

Optional Investments

	Description	Business Process Reengineering	Information Technology	Organizational Development	Public Communications	Staff Training	Cross Operating Unit Coordination
1							
2							
3							
4							
5							

DISINVESTMENTS (REDUCTIONS IN EXPENDITURES)

	Description	Business Process Reengineering	Information Technology	Organizational Development	Public Communications	Staff Training	Cross Operating Unit Coordination
1							
2							
3							
4							
5							

HONOLULU BOARD OF WATER SUPPLY					
STRATEGIC BUSINESS PLAN					
OPERATING UNIT PLAN					
Operating Unit: 0					
Strategic Investment Budget					
Capital Investments					
	Year 1	Year 2	Year 3	Year 4	Year 5
Capital Project #1	$0	$0	$0	$0	$0
Supporting Capital Project #1-1	$0	$0	$0	$0	$0
Supporting Capital Project #1-2	$0	$0	$0	$0	$0
Capital Project #2	$0	$0	$0	$0	$0
Capital Project #3	$0	$0	$0	$0	$0
Supporting Capital Project #3-1	$0	$0	$0	$0	$0
Supporting Capital Project #3-2	$0	$0	$0	$0	$0
Capital Project #4	$0	$0	$0	$0	$0
Total Capital Investments	$0	$0	$0	$0	$0
Operations & Maintenance Expenses					
	Year 1	Year 2	Year 3	Year 4	Year 5
Sponsoring Operating Unit Expenditure Projection					
Staffing Requirements (No. of FTE Positions)	***0***	***0***	***0***	***0***	***0***
Personnel	$0	$0	$0	$0	$0
Salaries & Wages	$0	$0	$0	$0	$0
Benefits	$0	$0	$0	$0	$0
Contractuals	$0	$0	$0	$0	$0
Commodities	$0	$0	$0	$0	$0
Commodity Category #1	$0	$0	$0	$0	$0
Commodity Category #2	$0	$0	$0	$0	$0
Non-CIP Capital	$0	$0	$0	$0	$0
Total Sponsoring OUL Base Expenses	$0	$0	$0	$0	$0
Other Operating Unit Expenditure Projection					
Total Sponsoring OUL Base Expenses	$0	$0	$0	$0	$0
Supporting Operations & Maintenance Expenditures					
Public Communications	$0	$0	$0	$0	$0
Stakeholder Involvement	$0	$0	$0	$0	$0
General Public Education	$0	$0	$0	$0	$0
Media Outreach	$0	$0	$0	$0	$0
Information Technology	$0	$0	$0	$0	$0
Hardware	$0	$0	$0	$0	$0
Network	$0	$0	$0	$0	$0
Software	$0	$0	$0	$0	$0

Staff Training				$0	$0	$0	$0	$0
	External Training			$0	$0	$0	$0	$0
	On-the-Job Training			$0	$0	$0	$0	$0
Total Supporting Expenses				$0	$0	$0	$0	$0
TOTAL OPERATIONS & MAINTENANCE EXPENSES				$0	$0	$0	$0	$0
TOTAL INVESTMENT EXPENSES				$0	$0	$0	$0	$0
					Discount Rate		**Net Present Value**	
NET PRESENT VALUE OF INVESTMENT EXPENSES					**6.00%**		**$0**	

Appendix C: Detailed Case Study and Glossary of Key Terms

"Development of a Strategic Business Plan – Portfolio Management for Public Utilities"

by Eric P. Rothstein and Donna Kiyosaki

NOTE:

The paper on the following pages provides more substantive detail on the analytical methods of the strategic planning process as described in Chapter 5 of this report – in particular, the identification, evaluation, and ranking of strategic investment options. The paper was developed, in part, as an interim deliverable for the Research Foundation project, ***Development of a Strategic Planning Process.*** The paper was presented at the AWWA 2002 National Conference in New Orleans, Louisiana, and is included in conference proceedings.

Subsequent to submittal of this paper to AWWA, ongoing project efforts resulted in further evolution of the strategic planning process; most notably, the final strategic planning process as presented earlier in the project final report has been condensed from eight steps to five. Information presented elsewhere in the final report represents the final results of the research project, while the paper included in this appendix represents the products of the research project in its earlier stages.

AwwaRF – Honolulu Board of Water Supply: Development of a Strategic Business Plan – Portfolio Management for Public Utilities

Eric Rothstein, Donna Kiyosaki

Executive Summary

The Honolulu Board of Water Supply (BWS) has engaged in a strategic business planning process to restructure the utility's organization, services, and community presence - and position the utility for success in the 21st century. Through support from the Awwa Research Foundation (AwwaRF), this tailored collaboration project will employ a set of processes and techniques that reflect the application of ***portfolio management theory*** to the challenge of strategic business planning in an increasingly complex and competitive utility market. The specific objectives of the AwwaRF-BWS project are to develop a strategic business plan for BWS, and a planning process, that demonstrates how water utilities can:

- Evaluate and develop market opportunities,
- Optimize asset use and investment,
- Evaluate and manage investment risks,
- Protect and expand market share, and perhaps most importantly,
- Enhance customer service and public perceptions.

The results of this project will be adapted into a guidance document for use by water utilities across North America, helping to advance the utility industry standard in strategic business management. This guidance document will build upon a breadth of literature and case studies that delineate the value of strategic business planning - particularly for organizations challenged by rapidly changing market conditions. For water utilities, this value is realized through:

- Alignment of major resource investments with strategic goals and objectives
- Focus on strategic (as opposed to tactical) investments and activities
- Effective balancing of multiple, often competing, objectives defined in both monetary and non-monetary terms
- Linkage of the strategic goal-setting process to annual budget and resource allocation processes
- Development of risk management strategies arising out of explicit consideration of operating and investment uncertainties
- Definition and tracking of utility performance through a limited number of objective, relevant metrics, directly tied to the organization's strategic objectives
- Development of utility management tools to facilitate strategic allocation of resources, sustain strategic business planning, and enable timely responses to changing market conditions

These outcomes may be secured through application of *portfolio management* – a well-established private-sector technique for investment decision-making – to the challenges of water utility management. Portfolio management contemplates a balancing of resource investment opportunities such that risks are managed and strategic objectives advanced. Its application to public-sector utilities invites development of business strategies that leverage market position and espouse entrepreneurial investment, once basic customer service requirements are addressed and available efficiencies of core services realized.

Integration of portfolio management-based strategic business planning is considered as a natural bridge between strategic goal-setting exercises and development of annual operating and capital improvement budgets. Procedurally, strategic goals and objectives are expressed in terms of a limited number of well-defined performance metrics. Strategic investments are then identified as those resource allocations (e.g., available funds, staff resources) that will yield substantial advances toward achievement of strategic goals – as defined by reference to the performance metrics associated with the strategic goals. To the extent practicable, each such investment is then fully characterized in terms of anticipated performance and associated risks. Potential investments are prioritized, based on projected performance, and incorporated into a strategic portfolio to achieve a balanced array of investments that mitigate uncertainties and are likely to realize potential returns. The selected portfolio is then reflected in the utility's subsequent budget and resource planning documents that focus more specifically on the tactical requirements of utility operations.

Strategic business planning is thereby a phase of the annual utility business planning cycle that provides the mechanism by which strategic goals are translated into a specific, priority work plan for the utility's operating units. Goals are established; performance metrics defined; strategic investments identified[1], evaluated, and prioritized; and selected investment portfolios budgeted. By focusing on those investments that are strategic in nature – those resource allocations that provide for substantive advance of established utility goals – utilities may avoid budgeting, by default, incremental adjustments to the status quo.

Strategic planning establishes an environment that supports and encourages innovation and challenges stagnation. Tactical details of line-item budgeting, business process re-engineering, and organizational restructuring are not contemplated as a means unto themselves but rather employed to effect strategic investments, as well as perpetuate sound, sustainable utility operations. Accordingly, a well-functioning strategic business planning process provides a 'living document' for effective utility management that is referred to regularly, updated without difficulty, and that compels focus on strategic performance measures. Long-term sustainability of the strategic business planning process may be supported through several techniques

---

[1] Notably, an environment should be constructed such that entrepreneurial strategies and strategic investments may be advanced and researched at any time in the business calendar. The Strategic Business Planning process may catalyze new strategy development or merely facilitate the evaluation of already identified strategic investments, while enabling subsequent allocation of resources in either event.

that ensure construction of an effective bridge between declarations of lofty strategic goals and line-item budgeting, including:

- Limitation of the number of strategic performance measures, to galvanize organizational focus
- Integration of strategic business planning into the annual business planning cycle
- Employ of concise, accessible, yet comprehensive tools for submittal of information on, and evaluation of, candidate strategic investment options
- Representation of the strategic investment portfolio in terms of 1-year and 5-year work programs (for which specific milestones and responsibilities are identified)

The Strategic Business Plan is thereby the principal vehicle by which utilities may assess their market position and define their responses to changing market conditions.

Introduction

Public water and wastewater utilities over the last decade have experienced a wholesale revision in the structure and dynamics of the marketplaces in which they operate. Regulatory requirements, system expansion demands, and limitations on federal and state infrastructure development support have imposed increasing pressures on revenue requirements. At the same time, private-sector competition and bundled service offerings have both galvanized demands for rate containment and modified expectations about utilities' traditional scopes of services. More than ever before, water utilities operate in a complex, competitive, and dynamic marketplace wherein survival is dependent on utilities' ability to protect and expand market share, manage risks, and improve customer satisfaction.

"Can we continue to run our business in the same way as we have in the past century? ... Not if we want to survive!"

-Cliff Jamile,
General Manager
Honolulu Board of
Water Supply

In this context, water utilities must evaluate their market position, act strategically, and deploy resources to respond to rapidly changing market conditions. As suggested by comparable utility services (e.g., telephone, electric, cable), success will be achieved through efficient delivery of traditional services in combination with new or expanded service offerings. Strategic business planning that acknowledges this dynamic context is required - suggesting application of techniques employed successfully in the private sector. In particular, the fundamental principles of ***portfolio management***, which has guided long-term growth and sustainability of some of the most successful business enterprises[2] and investment houses, prevail perhaps more strongly for water utilities.

A portfolio management perspective affords particular advantages for water utilities because of the heightened complexity of resource investment decisions. Whereas commercial interests will make investments through an assessment of the potential

[2] For example, Anheuser-Busch, Chase, Coca-Cola, GM, and the Southern Company, to name a few, are characterized by extensive, and actively managed, investment and new product portfolios.

monetary risks and returns of a portfolio in aggregate, water utilities face the more acute challenge of securing both monetary and non-monetary benefits where attendant risks are often more profound (e.g., public health, environmental degradation) and prospective returns are less certain or quantifiable. Water utilities must balance resource investments across multiple, competing objectives (e.g., environmental stewardship, support of economic development) to serve a plethora of stakeholder interests (e.g., development community, ratepayers, environmental interests, regulatory agencies). In these circumstances, application of portfolio management principles becomes even more compelling and will enable water utilities to:

- Optimize asset use and investment opportunities
- Fully leverage resources to enhance competitive success
- Effectively manage risks and conduct risk-based decision-making
- Effectively analyze monetary and non-monetary investment decisions within a single analytical framework
- Explicitly recognize benefits of asset allocation and risk diversification

The value of the AwwaRF-BWS tailored collaboration is to demonstrate the application of portfolio management principles to the development of a strategic business plan for the Honolulu BWS, outline the complications and benefits of this planning perspective, and illustrate how ***strategic*** business planning may be integrated into annual water utility planning processes which may (but often do not) link strategic goals to tactical resource allocations. In this way, the AwwaRF-BWS tailored collaboration will offer insights into how utility managers may more effectively anticipate and respond to changing water utility management challenges and market conditions.

Annual Business Planning Cycle

These changing market conditions suggest, more than ever before in the water utility industry, that business planning processes must ensure alignment of strategic goals with resource allocation decisions. For many water utilities, previously unchallenged by the need to develop a ***strategic*** business plan for monopolistic service delivery, this represents a significant paradigm shift. In any case, strategic business planning must become a regular, sustainable, component of the utility's existing planning processes to be an effective instrument for resource allocation decisions.

Notably, for many utilities, the connection between strategic goal setting and budget development is tenuous at best. Annual business planning processes commonly suffer from a disconnect between the policy orientation of strategic goal development and the tactical realities and minutia of line-item budgeting. Integrating a ***strategic*** business planning phase into the annual budget-setting process may serve to establish the missing linkage. Accordingly, the AwwaRF-BWS strategic business planning process is structured to leverage, rather than duplicate, a utility's typical planning activities and provides a bridge between the articulation of general strategic goals and objectives, and the utility's annual budget development process. It ensures the translation of utility-wide strategic goals into specific tactical plans and performance measures for which individual utility operating divisions are accountable. As such, it may be considered an intermediary step in the traditional water utility planning cycle.

By integrating the strategic business planning process into the overall utility planning cycle, and prescribing concise planning submittals that support budget development, the strategic planning process does not duplicate but rather supports budget development efforts. It is the strategic phase of the planning cycle that occurs in advance of the fundamentally tactical processes of budget development and organizational restructuring.

The specific steps of the AwwaRF-BWS strategic business planning process developed for this project (shown below) require utility management to assess their market position, identify strategic investment options, prioritize investments based on the extent to which strategic goals are advanced, and establish a portfolio of options for implementation. The strategic business planning process focuses specifically on those initiatives and investments that will effect a change in the status quo and allow the utility to advance its market position and meet any other strategic objectives. This process employs well-established evaluation techniques for quantifying monetary and non-monetary impacts, management of stakeholder involvement, and prioritization of strategic investment options[3]. As shown in the graphic below, the strategic business planning process maps to the typical utility business planning process, affording a mechanism to move from goal setting to line-item budgeting.

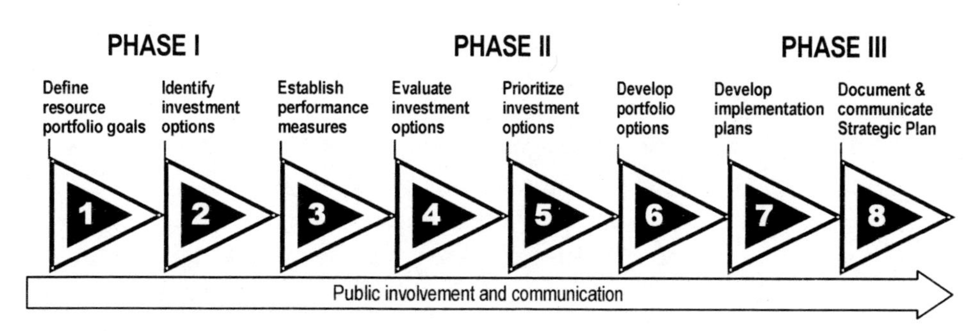

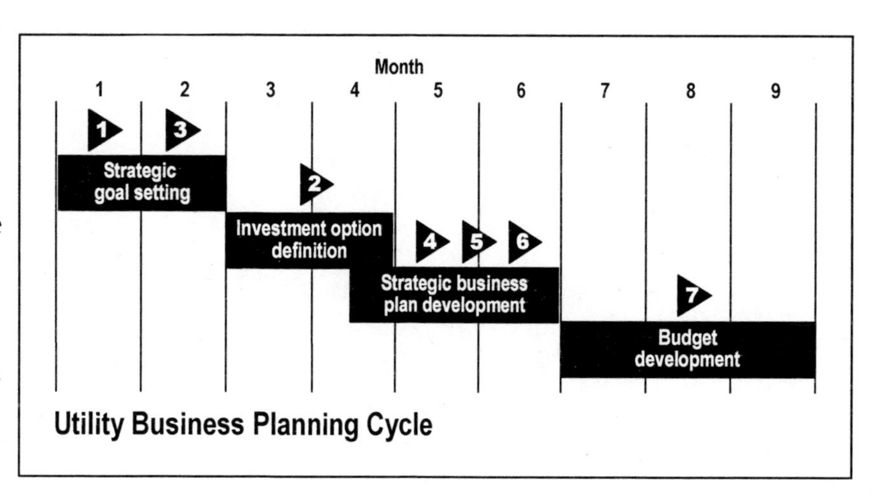

The available, and convenient, integration of these ***strategic*** business planning steps into the utility's annual planning cycle is perhaps the single most important factor for ensuring the long-term sustainability of a strategic plan. Products of the strategic business planning process offer concise utility

---

[3] For a fuller description of each step of the Strategic Business Planning process, see the AwwaRF-BWS project description titled: '***Development of a Strategic Business Plan'***; for more information on strategic investment portfolio construction, see the project briefing paper ***'Investment Options Evaluation Techniques.'***

management tools to ensure that strategic goals are advanced and necessary resource allocations prioritized. In particular, the strategic business planning process contemplates a planning phase in which utility management defines, in concise and measurable terms, the near-term and 5-year work program it will conduct to achieve (or advance toward) the utility's stated strategic goals. This work program will reflect a rigorous prioritization of strategic investment options - conducted in advance of budget submittals. It will serve as a management tool to monitor strategically important initiatives, and be readily updated or revised to effect responses to changing market conditions or organizational objectives.[4]

This strategic business planning process thereby addresses several opportunities for improvement of annual business planning cycles that prevail in many utilities. The process:

- ensures alignment of strategic goals and objectives with divisional business plans and budgets,
- effectively prioritizes investment options, and
- translates strategic goals to specific work assignments with defined performance measures and scheduled milestones.

> "To see victory only when it is within the ken of the common herd is not the acme of excellence."
>
> - *'Sun Tzu & The Art of War As Applied to Portfolio and Risk Management'*

Strategic Investments Options

"To see victory only when it is within the ken of the common herd is not the acme of excellence." While the strategic business planning process supports budget development, it is important to maintain focus on those investments and activities that are strategic in nature. By definition, line-item budgeting must contemplate the minutia of operating units' requirements for office supplies as well as personnel additions and technology resource acquisitions. Strategic investments may range from the employment and training of staff to increase intellectual capital, to investment in new facilities or products to create or capture new markets - but they are distinguished by one or more of the following criteria:

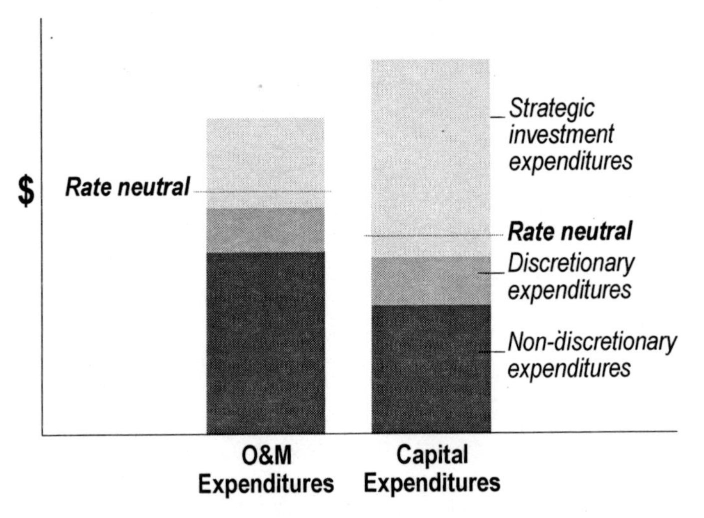

a. Create a new customer service that enhances value to the customer.
b. Fundamentally change the way a utility delivers a customer service, thereby enhancing value to the customer.

[4] Specific examples of templates used to develop components of the BWS Strategic Business Plan are provided as Appendix B of this report. These individual operating unit submittals may be readily revisited and updated; milestones and performance measures related to each operating unity will be summarized in an enterprise-wide schedule of strategic initiative milestones.

c. Provide a mechanism for managing a utility's monetary and non-monetary risks.

Perhaps most importantly, strategic investments result in measurable progress toward achievement of a utility's strategic goals.

Strategic Investments – Funding

For purposes of development of a strategic business plan, strategic investments are considered apart from typical annual capital and operating expenditures - the non-discretionary expenditures that comprise the basic, unalterable costs of performing core utility functions, and those discretionary expenditures that utilities incur to address community values irrespective of their strategic value. In some cases, strategic investments may be made by an allocation of resources within the utility's existing revenue requirements - and thereby be 'rate neutral'; in other cases strategic investments may require additional revenues. In many cases, strategic investments will require a multi-year allocation of resources, frequently from both capital and operating budgets. In any case, the utility business planning cycle contemplates the development (or update) of its strategic business plan in advance of the annual budgeting process. Accordingly, rate adjustments or reallocations of utility operating and capital budgets to serve the utility's strategic goals and objectives are defined in advance of annual budget development.

Strategic Investments – Identification

By definition, strategic investments are anticipated to convey benefits that overwhelm the discordance that typically accompanies such rate or budget decisions. They are unusual, represent change, and so require strong utility-wide commitment and perseverance. In many instances, their genesis will be a product of ongoing utility planning and re-engineering efforts; in other cases, investment opportunities may arise outside of typical utility planning processes. For example, strategic investments in new water supplies - including, in BWS's case, reclaimed water and desalination - would typically be identified and characterized through long-range water resource planning functions. However, atypical investments, like the potential use of abandoned lines for extension of fiber optic networks, may be brought forth through a plethora of external or anomalous mechanisms including vendors, community interests, and industry information exchanges (e.g., conferences, Board memberships). [5] One of the

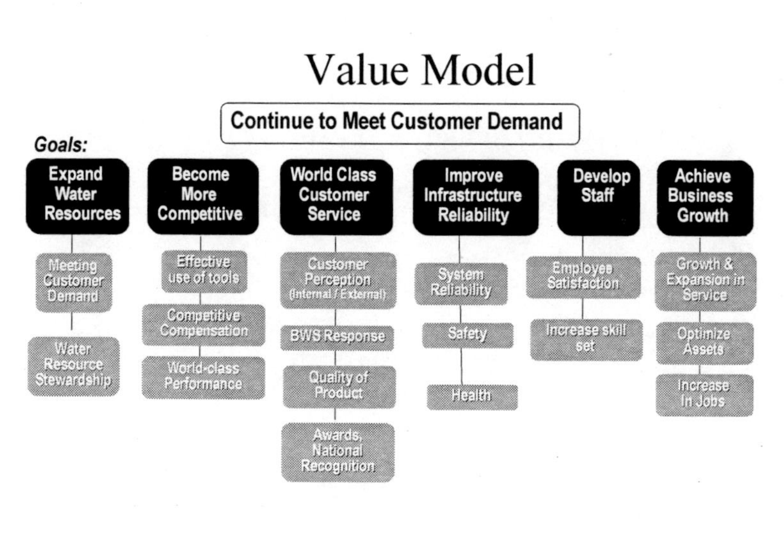

---

[5] In this respect, BWS's recent reorganization reflects a progressive response to the changing utility market and BWS's strategic goals. BWS has established a 'Business Development' operating unit charged with responsibility to 'explore and pursue new business opportunities for the benefit of the Board's present and future customers.'

benefits of an *institutionalized* strategic planning process is that it provides an established vehicle for consideration of strategic investment opportunities - a vehicle that may capture both creative ideas within traditional planning frameworks and embrace 'out-of-the-box' thinking from a broad range of stakeholder interests.

Strategic Investments – Performance Measurement

The rigor of an established strategic planning process (as opposed to reliance on the omniscience of executive leadership) imposes a requirement for structured evaluation of strategic investments through objective measurement of prospective performance. Strategic planning inherently demands selecting courses of action - investments - among a palette of available alternatives. Discerning the merits of one alternative relative to another (and defending one's assessment under the gaze of stakeholder scrutiny) is most effectively achieved by reference to measures of investment performance.

In contrast to the private sector's focus on monetary returns, investment performance among water utilities is a function of multiple goals, reflecting a broad diversity of community values. Water utilities are challenged with not only efficient, cost-effective, reliable service delivery, but also protection of public health and environmental resources, community service, and so on. For example, BWS's value model[6] is presented here and reflects the utility's intent to balance culturally significant resources with community economic development.

Measures of investment performance that reflect these multiple objectives are the best vehicles by which to evaluate and prioritize strategic investment options. Performance measures gauge the contributions of each utility investment to strategic goals in discrete, measurable terms. Characteristics of sound performance measures are precision, scalability, and substantive content. Importantly, performance measures must be constructed to facilitate rating of utility investments such that one investment option is determined to yield greater or less benefit than another option.

In general, performance evaluation may rely on two forms of measure:

- *Natural Scales* - wherein the extent of contribution to strategic goals is directly monitored. For example, unit costs per employee for various breakdowns of service delivery (e.g., million gallons sold, plant capacity) offer natural scales, whereby lower unit costs are 'better than' higher unit costs by the measured variance in cost levels.
- *Constructed Scales* - wherein the extent of contribution is defined by reference to an amalgam of conditions, activity levels, and directly measured impacts that delineate higher or lower performance. Narrative descriptions of performance are drafted to define what constitutes achievement of higher or lower levels of performance.

Readily collected natural measures (typically relating to unit costs or time requirements for task completion) are rarely definitive in terms of measuring a utility's achievements relative to its strategic goals. In any event, use of performance

6 'Value model' in this context, is imply a graphical representation of an organization's fundamental goals and strategic objectives.

measures that are aligned with the utility's strategic goals and objectives is critical for strategic planning to be effective. Utilities that maintain focus on a few, selected performance measures that reflect the primary mission of the organization are typically well-suited for a competitive, rapidly changing marketplace.

Strategic Investments – Risk Evaluation

In addition, the strategic planning process provides an opportunity to consciously and explicitly evaluate risks associated with each investment option and develop risk management strategies that will most effectively advance the utility's strategic goals.

Risk is the possibility of suffering harm or loss[7] – an undesirable outcome. Evaluation of risk requires assessment of the probabilities of occurrence, opportunities to reduce these probabilities, potential consequences of the risks occurring, and perceptions of damage. For water utilities charged with the protection of public health, certain investments are required to provide an exceptionally high level of protection from certain risks. However, as utilities face a changing marketplace, it is becoming increasingly important for utilities to distinguish those risks that require avoidance versus those that may be managed. Strategic planning is, in large measure, an exercise in risk management in the pursuit of strategic goals.

Risk is managed through reduction and allocation. Risk can be reduced by investing in measures that will minimize the probabilities of occurrence. Risk is allocated by assigning the risk, in whole or in part, to another party. Strategic investments have inherent risks, and different risk and potential benefit (or 'return') profiles. Strategic planning requires selection of strategic investments to achieve an appropriate balancing of risk (harm or loss) and contribution to achievement of strategic goals.

"In the water industry, we tend to practice risk avoidance rather than risk management. We need to examine risk carefully, especially when there are opportunities involved, and not always avoid them."

- John Huber, Chief Executive Officer
Louisville Water Company
August 10, 2001

Portfolio Management / Goals Achievement Evaluation

The concept of balancing risk and return is at the center of portfolio management theory, which has been applied successfully among investment fund managers and private industry executives. This theory postulates, when applied to water utilities, that a strategic approach to resource investment does not require each investment to be devoid of substantial risks. Rather, utilities may invest resources and engage in activities that individually represent a broad spectrum of risk and return tradeoff. Portfolio management requires that, in combination, these discrete investments reflect a balancing of risk and return that is consistent with the utility's strategic goals and objectives.

The fundamental steps of the finance-borne portfolio management approach are relatively simple, straight-forward, and applicable to the water utility

[7] *The American Heritage Dictionary of the English Language* (American Heritage Publishing Co., Inc. – 1976).

context.[8] They involve determining the set of investments available to the water utility organization and their characteristics in terms of potential benefits, risks, and resource requirements. Investment opportunities are then prioritized within and across functional areas. Finally, investment options are selected for inclusion in the utility's strategic business plan by virtue of their alignment with established organizational goals and objectives.

Prioritization of investment options may be conducted through construction of an analytical model that quantifies the relative importance of strategic goals, calculates total benefits associated with individual investments, and evaluates potential changes in investment rankings given alternative values of input variables. This formal prioritization may be conducted using **multi-attribute utility analysis (MUA)**, which has been applied successfully for the AwwaRF Public-Private Partnership Evaluation and Capital Planning Strategy projects. The significant benefit of MUA, especially applied within a public utility context, is the ability to incorporate consideration of both non-monetary and monetary impacts within a single analytical framework.

The analytical structure of MUA is predicated on the development of independent, non-redundant, and comprehensive fundamental objectives for the prioritization decision. Accordingly, portfolio construction is best served when a water utility's strategic goals have these characteristics.[9] Once established, the objectives are weighted based on their relative importance using a step-wise process that considers the merit of tradeoffs of incremental improvement in the accomplishment of one objective at the cost of an incremental decline in the achievement of another objective.

"MUA is actually not complicated at all; it is surprisingly intuitive and very powerful for a whole host of applications."

- Bevin Beaudet,
Senior Vice President
CH2M HILL
April 2001

With criteria weights established, investment option performance is scored using either constructed or natural scales, which define the benefits accruing from given investments. Whether natural or constructed, performance scales are used to 'score' investment option performance and indicate the benefits in terms of accomplishment of strategic goals. Prioritization of investment options is an arithmetic exercise once criteria weights have been assigned to fundamental objectives and investment option performance scored. Total benefits of each option are calculated and options are ranked on the basis of either total benefit or benefit-to-cost ratio.

---

8 These steps, in terms of financial portfolio management include: (1) valuation-describing a universe of assets in terms of expected return and expected risk; (2) asset allocation decision-determining how assets are to be distributed among classes of investment, such as stocks or bonds; (3) portfolio optimization-reconciling risk and return in selecting the securities to be included, such as determining which portfolio of stocks offers the best return for a given level of expected risk; (4) performance measurement-dividing each stock's performance (risk) into market-related (systematic) and industry/security-related (residual) classifications. – *Barron's Finance and Investment Handbook*

9 The Honolulu BWS's current strategic goals were developed in advance of a strategic planning process wherein formal prioritization analysis is used to evaluate investment options and construct an investment 'portfolio'. See BWS strategic goals presented at July 2001 Board meeting.

Investment portfolios are constructed to yield the greatest benefit (or 'returns' in both monetary and non-monetary terms) by examining the prevailing risks associated with investment options. Risk evaluation may be reflected either directly, through the construction of risk-sensitive performance measures, or indirectly through sensitivity analyses of the initial investment prioritization.

Scenario Planning

Risk management may be facilitated through exercises in scenario planning, which can help strategic planners:

- enhance the accuracy of assigned probabilities that risks will occur,
- identify opportunities for reducing risks, and
- delineate the consequences of risks occurring.

Scenario planning characterizes the business environment within which a utility will most likely operate over time and potential favorable or unfavorable changes that may affect business performance. Several scenarios may be developed to represent worst, expected, and best case conditions. By examining potential changes to its business environment, a utility may articulate a strategic direction and define risk-management methods that will buffer the organization against adverse business environment changes. Indicators or "triggers" that forewarn the utility of a change in the business environment may be identified so that steps can be taken to mitigate potential adverse consequences or take advantage of emerging opportunities.

Scenario planning techniques may be used at multiple points in the strategic business planning process - as part of the development of strategic goals and objectives for the organization, as part of the evaluation of individual investment options, and as part of the construction of investment portfolios. To the extent that strategic planning is an ongoing, iterative process, scenario planning provides a foundation for evaluation and re-evaluation of investment decisions. Fundamentally, it provides a convenient, accessible vehicle to place into context the strategic decision-making required by effective portfolio management.

Integration of Strategic Business Planning into Annual Planning Cycle

The iterative nature of strategic business planning, and its focus on strategic investments as opposed to tactical activities, portend a potential danger. The process may become so laborious and so isolated from daily operational challenges that its relevance and effectiveness in advancing the utility's strategic goals is compromised.

"It's an evolutionary versus a revolutionary process. It requires achievable steps that people can adapt to within a controlled environment, without creating anarchy."
- John Huber, Chief Executive Officer
Louisville Water Company
August 2001

The challenge of strategic planning is to effect translation of strategic goals into utility accomplishments, to enhance utility preparedness, and to maintain utility focus in an ever-changing marketplace. Accordingly, strategic business planning will be most effective as an integral component of the annual utility planning cycle as opposed to an irregular, onerous exercise. In large measure, strategic business planning serves to bridge the gap between development of strategic goals and planning for implementation of specific tactical measures (e.g., budget development, business process re-engineering, organizational development).

A variety of documentation instruments may be employed to construct this planning bridge, but each must serve several fundamental purposes to be effective. The annual strategic plan must:

- establish performance measures associated with the utility's strategic goals,
- identify specific, near-term objectives that contribute to accomplishment of strategic goals (as indicated by reference to performance measures), and
- identify resource requirements and milestones for accomplishment of planned work.

In summary, annually updated strategic business plans codify managerial responsibilities of the utility's executive leadership team. They establish the priorities for which managers will be held accountable, as well as the means by which accomplishment will be measured, and they ensure that milestones are aligned with the utility's overall strategic goals. Brevity and specificity are the hallmarks of sustainability.[10] Further, in order to be integrated within the annual planning cycle, strategic planning must be completed between the annual review and revision of strategic goals and the budget development process - typically a 2- to 4-month period.

Strategic business planning is distinguished from, and provides a foundation for, detailed operating unit (e.g., divisional) business plans. In particular, strategic planning focuses on the accomplishment of strategic goals, as indicated by reference to a limited set of performance measures. Operating unit business plans extend beyond these requirements to address the tactical requirements for operational excellence.

Business Plan Evolution

The annual business planning cycle may therefore be described as an incremental process of constructing a well-defined framework for utility performance management. The cycle begins with definition (or revision) of strategic goals, and the strategic business plan subsequently focuses on the identification, evaluation, selection, and planning of strategic investments; the cycle then moves to the development of tactical actions. For the Strategic Planning increment, portfolio management approaches may be used to ensure optimal resource allocation. Once done, the utility's Strategic Plan is incorporated into each operating unit's business plan to address tactical requirements. Each planning phase or instrument has common elements - milestones, benchmarks and performance measures, budget estimates - but their foci and level of detail are importantly distinct.

---

[10] Accordingly, for BWS, Operating Unit Leads (OULs) completed a brief business planning template that requires the above listed information through 1-2 hour interview sessions.

Strategic Business Plan

Strategic business planning, by definition, addresses the utility's 'change initiatives' - those efforts intended to enhance or otherwise modify the scope of services. Effective strategic business plans identify a limited set of strategic investments on which the organization will focus - thereby avoiding frenetic activities or re-engineering for its own sake - and outline the utility's risk management strategy. Appropriate benchmarks and performance measures referenced for strategic planning will typically relate to broad indicators of overall efficiency and effectiveness (e.g., O&M cost/mgd treated, debt/equity ratios, customer satisfaction rating, average age of built infrastructure, and order-of-magnitude capital and operating cost projections). Milestones will typically reflect the long-term nature of strategic investments, defining key steps toward achievement of 'best-in-class' performance goals.

Operating Unit Business Plans

Building on the strategic plan, individual operating unit business planning will offer substantially greater transactional detail on *how* strategic investments are to be accomplished, as well as daily operating requirements to be implemented. These plans will define the operating units' responsibilities relative to the organization's strategic goals and outline activities that, while not necessarily strategic in nature, sustain customer service and financial integrity. Appropriate benchmarks and performance measures will typically relate to specific aspects of each operating unit's scope of responsibilities and be transactional in nature (e.g., customer billings per month, number of new wells developed with a defined sustainable yield, percentage reduction in documented backlog of line replacements). Milestones will typically define relatively short-term measures to enhance effectiveness and efficiency (e.g., extend Automatic Meter Reading to additional portions of service area, complete 3-D hydraulic modeling of system, develop reclaimed water contract instrument). These plans serve as the basis for line-item operating and capital budgets; they may also address requirements for specific business process revisions and staffing re-alignments.

Annual Review and Update

By integrating strategic planning into the annual utility planning cycle, a focus is maintained on accomplishment of strategic goals and responsiveness to changing market conditions. Sustainability is further assured by recognizing that revision and editing almost always require less time and effort than original authorship. Within an annual planning cycle, the strategic planning process amounts to:

- Re-examination and revision of the utility's strategic goals and key performance measures
- Re-evaluation and confirmation of risks and risk management strategies (through updated scenario planning)
- Review and update of 1st and 5-year milestones for achievement of strategic goals
- Review of implications of updates for tactical operating unit business plans (and requirements for cross-operating-unit cooperation to achieve synergies)

- and may be documented through limited revisions to the utility's then-current Strategic Plan.

Conclusions

Strategic business planning is optimally considered an integral step in the annual utility planning cycle. It provides a bridge between the establishment of strategic goals and the tactical focus of line-item budgeting and business process re-engineering. To be effective, strategic planning will enforce a focus on individual operating units' contributions to utility-wide strategic goals, thereby ensuring alignment of strategic goals with near-term operating unit milestones and budgets. Ideally, strategic business planning forms a bridge between the annual budgeting process and the ongoing planning processes within a utility.

Portfolio management approaches may be used as a tool for managing strategic investment selection and risk management. It is particularly well-suited to the emerging challenges of the water utility industry as it accommodates and highlights utilities' responsibilities for balancing multiple, often competing goals in a setting replete with public scrutiny.

Strategic planning activities need not be protracted, and *must* not be onerous, to remain sustainable. Rather, strategic planning may be conducted through a series of simple steps that define, in practical terms, how and at what pace the utility will advance toward accomplishment of its strategic goals. In so doing, it creates a foundation for operating unit business planning and line-item budgeting that will help ensure these activities serve, rather than dictate, the utility's positioning in an increasingly competitive marketplace.

References

Sun Tzu & The Art of War As Applied to Portfolio and Risk Management by Jim Hight, Strategies&Tactics.com.

The Fall and Rise of Strategic Planning by Henry Mintzberg, Harvard Business Review, January - February 1994, pp. 107-114.

Project Portfolio Management White Paper by InfoHarvest, Inc.

Glossary of Key Terms

Budget and Resource Planning - Essentially tactical manifestations of the annual strategic business planning process involving budgeting and staffing analyses of strategic investment resource requirements.

Multi-Attribute Utility Analysis (MUA) – A formal decision analysis technique that incorporates consideration of non-monetary and monetary impacts within a single analytical framework, wherein the relative importance of strategic goals are articulated, performance metrics relative to each goal established, and total benefits associated with individual investments estimated based on the established goals and performance metrics.

Portfolio Management – A strategic planning theory aimed at achieving a balanced array of investments that will mitigating uncertainties and maximize the probability of achievement of potential returns.

Scenario Planning – A strategic planning tool involving the characterization of the business environment within which a utility will most likely operate over time as well as potential favorable and unfavorable changes that may affect business performance. Used for the development of a business strategy affording appropriate consideration of prevailing risks.

Strategic Business Plan – The principle vehicle by which utilities may assess their market position and define their response to market conditions; a bridge between strategic goal-setting and tactical development of annual capital and operating budgets.

Strategic Business Planning – A phase of the annual utility planning cycle that provides a mechanism by which strategic goals are translated into specific, prioritized work plans for translation into tactical activity by the utility's operating units.

Strategic Goals – The set of specific utility performance attributes that define the utility's vision of its desired future and may be used to galvanize organizational focus.

Strategic Investments – Those resource allocations(e.g., available funds, staff resources) that will yield substantial advances toward achievement of the utility's strategic goals. Strategic investments include resource allocations that: (1) create a new customer service that enhances value to the customer; (2) fundamentally change the way a utility delivers a customer service, thereby enhancing value; or (3) provide a mechanism for managing a utility's monetary and non-monetary risks.

Strategic Objectives – a limited number of well-defined performance metrics used to indicate relative degrees of accomplishment of the utility's strategic goals.

Tactical Planning – Specific resource allocations or re-allocations including line-item budgeting, business process reengineering, and organizational restructuring oriented toward affecting strategic investments but also ongoing utility operations, maintenance, and capital investment processes

Appendix D: Bibliography

Bibliography

CH2M HILL. 2001. *Capital Planning Strategy Manual* [CD-ROM]. Denver, Colo.: AwwaRF and AWWA.

Clemen, Robert T. 1996. *Making Hard Decisions: An Introduction to Decision Analysis.* Belmont, Calif.: Duxbury Press.

Dawn of the Replacement Era: Reinvesting in Drinking Water Infrastructure. May 2001. Denver, Colo.: AWWA.

Edwards, E. 1997. How to Use Multiattribute Utility Theory for Social Decision Making. *IEEE Trans. Systems Man., Cybern.* 7:326–340.

Hight, Jim. Sun Tzu & The Art of War As Applied to Portfolio and Risk Management [Online]. Available: <Strategies&Tactics.com>.

Mintzberg, Henry. Jan.–Feb. 1994. The Fall and Rise of Strategic Planning. *Harvard Business Review,* pp. 107–114.

Project Portfolio Management—White Paper. InfoHarvest, Inc.

Saaty, Thomas L. 1992. *Decision Making for Leaders.* Pittsburgh: RWS Publications.

Strategic Planning for Public and Nonprofit Organizations: A Guide to Strengthening and Sustaining Organization Achievement. 1988. Jossey-Bass.

Von Winterfelt, D., and W. Edwards. 1986. *Decision Analysis and Behavioral Research.* Cambridge University Press.